Flinn Scientific
ChemTopic™ Labs

Kinetics

Senior Editor

Irene Cesa
Flinn Scientific, Inc.
Batavia, IL

Curriculum Advisory Board

Bob Becker
Kirkwood High School
Kirkwood, MO

Kathleen J. Dombrink
McCluer North High School
Florissant, MO

Robert Lewis
Downers Grove North High School
Downers Grove, IL

John G. Little
St. Mary's High School
Stockton, CA

Lee Marek
Naperville North High School
Naperville, IL

John Mauch
Braintree High School
Braintree, MA

Dave Tanis
Grand Valley State University
Allendale, MI

FLINN SCIENTIFIC INC.
"Your Safer Source for Science Supplies"
P.O. Box 219 • Batavia, IL 60510
1-800-452-1261 • www.flinnsci.com

ISBN 1-877991-82-1

Copyright © 2003 Flinn Scientific, Inc.

All rights reserved. No part of this book may be reproduced or transmitted in any form or by any means,
electronic or mechanical, including, but not limited to photocopy, recording, or any information
storage and retrieval system, without permission in writing from Flinn Scientific, Inc.
No part of this book may be included on any Web site.

Reproduction permission is granted only to the science teacher who has purchased this volume of Flinn
ChemTopic™ Labs, Kinetics, Catalog No. AP6369 from Flinn Scientific, Inc. Science teachers
may make copies of the reproducible student pages for use only by their students.

Printed in the United States of America.

Table of Contents

	Page
Flinn ChemTopic™ Labs Series Preface	i
About the Curriculum Advisory Board	ii
Kinetics Preface	iii
Format and Features	iv–v
Experiment Summaries and Concepts	vi–vii

Experiments

Introduction to Reaction Rates	1
Temperature and Reaction Rates	13
The Order of Reaction	23
Kinetics of Dye Fading	35
Determining a Rate Law	47

Demonstrations

Iodine Clock Reaction	61
Now You See It—Now You Don't	65
Sudsy Kinetics	69
The Pink Catalyst	72
The Floating Catalyst	75

Supplementary Information

Safety and Disposal Guidelines	78
National Science Education Standards	80
Master Materials Guide	82

Flinn ChemTopic™ Labs Series Preface
Lab Manuals Organized Around Key Content Areas in Chemistry

In conversations with chemistry teachers across the country, we have heard a common concern. Teachers are frustrated with their current lab manuals, with experiments that are poorly designed and don't teach core concepts, with procedures that are rigid and inflexible and don't work. Teachers want greater flexibility in their choice of lab activities. As we further listened to experienced master teachers who regularly lead workshops and training seminars, another theme emerged. Master teachers mostly rely on collections of experiments and demonstrations they have put together themselves over the years. Some activities have been passed on like cherished family recipe cards from one teacher to another. Others have been adapted from one format to another to take advantage of new trends in microscale equipment and procedures, technology innovations, and discovery-based learning theory. In all cases the experiments and demonstrations have been fine-tuned based on real classroom experience.

Flinn Scientific has developed a series of lab manuals based on these "cherished recipe cards" of master teachers with proven excellence in both teaching students and training teachers. Created under the direction of an Advisory Board of award-winning chemistry teachers, each lab manual in the Flinn ChemTopic™ Labs series contains 4–6 student-tested experiments that focus on essential concepts and applications in a single content area. Each lab manual also contains 4–6 demonstrations that can be used to illustrate a chemical property, reaction, or relationship and will capture your students' attention. The experiments and demonstrations in the Flinn ChemTopic™ Labs series are enjoyable, highly focused, and will give students a real sense of accomplishment.

Laboratory experiments allow students to experience chemistry by doing chemistry. Experiments have been selected to provide students with a crystal-clear understanding of chemistry concepts and encourage students to think about these concepts critically and analytically. Well-written procedures are guaranteed to work. Reproducible data tables teach students how to organize their data so it is easily analyzed. Comprehensive teacher notes include a master materials list, solution preparation guide, complete sample data, and answers to all questions. Detailed lab hints and teaching tips show you how to conduct the experiment in your lab setting and how to identify student errors and misconceptions before students are led astray.

Chemical demonstrations provide another teaching tool for seeing chemistry in action. Because they are both visual and interactive, demonstrations allow teachers to take students on a journey of observation and understanding. Demonstrations provide additional resources to develop central themes and to magnify the power of observation in the classroom. Demonstrations using discrepant events challenge student misconceptions that must be broken down before new concepts can be learned. Use demonstrations to introduce new ideas, illustrate abstract concepts that cannot be covered in lab experiments, and provide a spark of excitement that will capture student interest and attention.

Safety, flexibility, and choice

Safety always comes first. Depend on Flinn Scientific to give you upfront advice and guidance on all safety and disposal issues. Each activity begins with a description of the hazards involved and the necessary safety precautions to avoid exposure to these hazards. Additional safety, handling, and disposal information is also contained in the teacher notes.

The selection of experiments and demonstrations in each Flinn ChemTopic™ Labs manual gives you the flexibility to choose activities that match the concepts your students need to learn. No single teacher will do all of the experiments and demonstrations with a single class. Some experiments and demonstrations may be more helpful with a beginning-level class, while others may be more suitable with an honors class. All of the experiments and demonstrations have been keyed to national content standards in science education.

Chemistry is an experimental science!

Whether they are practicing key measurement skills or searching for trends in the chemical properties of substances, all students will benefit from the opportunity to discover chemistry by doing chemistry. No matter what chemistry textbook you use in the classroom, Flinn ChemTopic™ Labs will help you give your students the necessary knowledge, skills, attitudes, and values to be successful in chemistry.

About the Curriculum Advisory Board

Flinn Scientific is honored to work with an outstanding group of dedicated chemistry teachers. The members of the Flinn ChemTopic Labs Advisory Board have generously contributed their proven experiments, demonstrations, and teaching tips to create these topic lab manuals. The wisdom, experience, creativity, and insight reflected in their lab activities guarantee that students who perform them will be more successful in learning chemistry. On behalf of all chemistry teachers, we thank the Advisory Board members for their service to the teaching profession and their dedication to the field of chemistry education.

Bob Becker teaches chemistry and AP chemistry at Kirkwood High School in Kirkwood, MO. Bob received his B.A. from Yale University and M.Ed. from Washington University and has 16 years of teaching experience. A well-known demonstrator, Bob has conducted more than 100 demonstration workshops across the U.S. and Canada and is currently a Team Leader for the Flinn Foundation Summer Workshop Program. His creative and unusual demonstrations have been published in the *Journal of Chemical Education,* the *Science Teacher,* and *Chem13 News*. Bob is the author of two books of chemical demonstrations, *Twenty Demonstrations Guaranteed to Knock Your Socks Off, Volumes I and II,* published by Flinn Scientific. Bob has been awarded the James Bryant Conant Award in High School Teaching from the American Chemical Society, the Regional Catalyst Award from the Chemical Manufacturers Association, and the Tandy Technology Scholar Award.

Kathleen J. Dombrink teaches chemistry and advanced-credit college chemistry at McCluer North High School in Florissant, MO. Kathleen received her B.A. in Chemistry from Holy Names College and M.S. in Chemistry from St. Louis University and has more than 31 years of teaching experience. Recognized for her strong support of professional development, Kathleen has been selected to participate in the Fulbright Memorial Fund Teacher Program in Japan and NEWMAST and Dow/NSTA Workshops. She served as co-editor of the inaugural issues of *Chem Matters* and was a Woodrow Wilson National Fellowship Foundation Chemistry Team Member for more than 11 years. Kathleen is currently a Team Leader for the Flinn Foundation Summer Workshop Program. Kathleen has received the Presidential Award, the Midwest Regional Teaching Award from the American Chemical Society, the Tandy Technology Scholar Award, and a Regional Catalyst Award from the Chemical Manufacturers Association.

Robert Lewis teaches chemistry and AP chemistry at Downers Grove North High School in Downers Grove, IL. Robert received his B.A. from North Central College and M.A. from University of the South and has more than 26 years of teaching experience. He was a founding member of Weird Science, a group of chemistry teachers that has traveled throughout the country to stimulate teacher interest and enthusiasm for using demonstrations to teach science. Robert was a Chemistry Team Leader for the Woodrow Wilson National Fellowship Foundation and is currently a Team Leader for the Flinn Foundation Summer Workshop Program. Robert has received the Presidential Award, the James Bryant Conant Award in High School Teaching from the American Chemical Society, the Tandy Technology Scholar Award, a Regional Catalyst Award from the Chemical Manufacturers Association, and a Golden Apple Award from the State of Illinois.

John G. Little teaches chemistry and AP chemistry at St. Mary's High School in Stockton, CA. John received his B.S. and M.S. in Chemistry from University of the Pacific and has more than 36 years of teaching experience. Highly respected for his well-designed labs, John is the author of two lab manuals, *Chemistry Microscale Laboratory Manual* (D.C. Heath), and *Microscale Experiments for General Chemistry* (with Kenneth Williamson, Houghton Mifflin). He is also a contributing author to *Science Explorer* (Prentice Hall) and *World of Chemistry* (McDougal Littell). John served as a Chemistry Team Leader for the Woodrow Wilson National Fellowship Foundation from 1988 to 1997 and is currently a Team Leader for the Flinn Foundation Summer Workshop Program. He has been recognized for his dedicated teaching with the Tandy Technology Scholar Award and the Regional Catalyst Award from the Chemical Manufacturers Association.

Lee Marek retired from teaching chemistry at Naperville North High School in Naperville, IL and currently works at the University of Illinois—Chicago. Lee received his B.S. in Chemical Engineering from the University of Illinois and M.S. degrees in both Physics and Chemistry from Roosevelt University. He has more than 31 years of teaching experience and is currently a Team Leader for the Flinn Foundation Summer Workshop Program. His students have won national recognition in the International Chemistry Olympiad, the Westinghouse Science Talent Search, and the Internet Science and Technology Fair. Lee was a founding member of ChemWest, a regional chemistry teachers alliance, and led this group for 15 years. Together with two other ChemWest members, Lee also founded Weird Science and has presented 500 demonstration and teaching workshops for more than 300,000 students and teachers across the country. Lee has performed science demonstrations on the *David Letterman Show* 20 times. Lee has received the Presidential Award, the James Bryant Conant Award in High School Teaching from the American Chemical Society, the National Catalyst Award from the Chemical Manufacturers Association, and the Tandy Technology Scholar Award.

John Mauch teaches chemistry and AP chemistry at Braintree High School in Braintree, MA. John received his B.A. in Chemistry from Whitworth College and M.A. in Curriculum and Education from Washington State University and has 26 years of teaching experience. John is an expert in "writing to learn" in the chemistry curriculum and in microscale chemistry. He is the author of two lab manuals, *Chemistry in Microscale, Volumes I and II* (Kendall/Hunt). He is also a dynamic and prolific demonstrator and workshop leader. John has presented the Flinn Scientific Chem Demo Extravaganza show at NSTA conventions for eight years and has conducted more than 100 workshops across the country. John was a Chemistry Team Member for the Woodrow Wilson National Fellowship Foundation program for four years and is currently a Board Member for the Flinn Foundation Summer Workshop Program.

Dave Tanis is Associate Professor of Chemistry at Grand Valley State University in Allendale, MI. Dave received his B.S. in Physics and Mathematics from Calvin College and M.S. in Chemistry from Case Western Reserve University. He taught high school chemistry for 26 years before joining the staff at Grand Valley State University to direct a coalition for improving pre-college math and science education. Dave later joined the faculty at Grand Valley State University and currently teaches courses for pre-service teachers. The author of two laboratory manuals, Dave acknowledges the influence of early encounters with Hubert Alyea, Marge Gardner, Henry Heikkinen, and Bassam Shakhashiri in stimulating his long-standing interest in chemical demonstrations and experiments. Continuing this tradition of mentorship, Dave has led more than 40 one-week institutes for chemistry teachers and served as a Team Member for the Woodrow Wilson National Fellowship Foundation for 13 years. He is currently a Board Member for the Flinn Foundation Summer Workshop Program. Dave received the College Science Teacher of the Year Award from the Michigan Science Teachers Association.

Preface
Kinetics

Many chemical reactions that students see in the lab—indicator color changes, formation of a precipitate, evolution of a gas—occur immediately upon mixing. The rate of the reaction appears to be controlled by diffusion. Other reactions, however, occur at a slower rate and students can follow the progress of the reactions over time. What factors determine how fast a chemical reaction will occur? The answer to this important question has applications not just in chemistry, but also in engineering, food science, and human physiology. Kinetics is the study of the rates of chemical reactions. Chemists must be able to measure and control reaction rates in order to make compounds both safely and economically. The purpose of *Kinetics,* Volume 14 in the Flinn ChemTopic™ Labs series, is to provide high school chemistry teachers with laboratory activities that will help students understand and apply the principles of kinetics. Five experiments and five demonstrations allow students to measure reaction rates and identify how and why reaction conditions affect reaction rates.

Introducing Kinetics

How can the rate of a chemical reaction be measured? What effect does changing the concentration of reactants or their temperature have on the rate of a chemical reaction? Two experiments offer different approaches to these fundamental questions. In "Introduction to Reaction Rates," students measure the time required for the "blue bottle" reaction of dextrose and methylene blue. The procedure is simple and easy-to-do, perfect for a hands-on introduction to a difficult topic. In "Temperature and Reaction Rates," an inquiry-based experiment, students must design and carry out a procedure to analyze the rate of reaction of magnesium with hydrochloric acid at different temperatures. Comprehensive teacher notes include suggestions for extending the activity to investigate how the nature of the reactants, their concentration, or the addition of a catalyst influences the reaction rate. Alternatively, teachers may choose to introduce the study of kinetics with a demonstration. Both the "Iodine Clock Reaction" and the "Sudsy Kinetics" demonstrations can be used to lay the foundation for higher level experiments dealing with reaction orders and rate laws.

Reaction Orders and the Rate Law

The mathematical relationship between the rate of a reaction and the concentration of reactants is expressed in the order of reaction for each reactant and the overall rate law for the reaction. In "The Order of Reaction," students determine the order of reaction for a microscale iodine clock reaction by measuring the rate of the reaction starting with different concentrations of a reactant. Teachers who have avoided traditional iodine clock experiments because of their complexity will love the simplicity of this microscale version. In this high-tech age, however, it is also important to demonstrate to students how scientists use technology in the real world to study these topics. In "Kinetics of Dye Fading," students use colorimetry and graphical analysis to investigate how the rate of fading of a familiar indicator depends on its concentration. Finally, "Determining a Rate Law" is a culminating-type activity—students apply what they have learned about reaction rates to evaluate the overall rate law for a reaction.

Catalysts and Reaction Pathways

As mentioned above, studying reaction rates is very important from a practical point of view. After all, if a reaction is too slow, it may not be practical. Too fast, however, and the reaction may not be safe! Kinetics is also important from a theoretical point of view in terms of understanding how reactions occur at the molecular level. Models of how reactions occur are called reaction pathways. In most textbooks, the reaction pathway is presented as an abstract concept and there are few suggestions for making it seem real to students. The demonstration "Now You See It, Now You Don't" gives teachers a springboard for introducing this abstract concept using the unusual, rhythmic color changes of an oscillating chemical reaction. The activity of catalysts in speeding up chemical reactions reflects their role in reaction pathways. Two demonstrations, "The Pink Catalyst" and "The Floating Catalyst," illustrate what catalysts do and how they work.

Safety, flexibility, and choice

Depend on Flinn ChemTopic™ Labs to give you the information and confidence you need to work safely with your students and help them succeed. As your safer source for science supplies, Flinn Scientific promises you the most complete, reliable, and practical safety information for every potential lab hazard. Each experiment and demonstration in *Kinetics* has been thoroughly tested and retested. You know they will work! At Flinn Scientific, we also know that no two classrooms are alike—the broad selection of experiments and demonstrations in *Kinetics* gives you the ability to design an effective lab curriculum for your students, with your resources, and to meet your local and state standards. Use the experiment summaries and concepts on the following pages to locate the concepts you want to teach and to choose activities that will help you meet your goals.

Format and Features

Flinn ChemTopic™ Labs

All experiments and demonstrations in Flinn ChemTopic™ Labs are printed in a 10⅞″ × 11″ format with a wide 2″ margin on the inside of each page. This reduces the printed area of each page to a standard 8½″ × 11″ format suitable for copying.

The wide margin assures you the entire printed area can be easily reproduced without hurting the binding. The margin also provides a convenient place for teachers to add their own notes.

Concepts — Use these bulleted lists along with state and local standards, lesson plans, and your textbook to identify activities that will allow you to accomplish specific learning goals and objectives.

Background — A balanced source of information for students to understand why they are doing an experiment, what they are doing, and the types of questions the activity is designed to answer. This section is not meant to be exhaustive or to replace the students' textbook, but rather to identify the core concepts that should be covered before starting the lab.

Experiment Overview — Clearly defines the purpose of each experiment and how students will achieve this goal. Performing an experiment without a purpose is like getting travel directions without knowing your destination. It doesn't work, especially if you run into a roadblock and need to take a detour!

Pre-Lab Questions — Making sure that students are prepared for lab is the single most important element of lab safety. Pre-lab questions introduce new ideas or concepts, review key calculations, and reinforce safety recommendations. The pre-lab questions may be assigned as homework in preparation for lab or they may be used as the basis of a cooperative class activity before lab.

Materials — Lists chemical names, formulas, and amounts for all reagents—along with specific glassware and equipment—needed to perform the experiment as written. The material dispensing area is a main source of student delay, congestion, and accidents. Three dispensing stations per room are optimum for a class of 24 students working in pairs. To safely substitute different items for any of the recommended materials, refer to the *Lab Hints* section in each experiment or demonstration.

Safety Precautions — Instruct and warn students of the hazards associated with the materials or procedure and give specific recommendations and precautions to protect students from these hazards. Please review this section with students before beginning each experiment.

Procedure — This section contains a stepwise, easy-to-follow procedure, where each step generally refers to one action item. Contains reminders about safety and recording data where appropriate. For inquiry-based experiments the procedure may restate the experiment objective and give general guidelines for accomplishing this goal.

Data Tables — Data tables are included for each experiment and are referred to in the procedure. These are provided for convenience and to teach students the importance of keeping their data organized in order to analyze it. To encourage more student involvement, many teachers prefer to have students prepare their own data tables. This is an excellent pre-lab preparation activity—it ensures that students have read the procedure and are prepared for lab.

Post-Lab Questions or Data Analysis — This section takes students step-by-step through what they did, what they observed, and what it means. Meaningful questions encourage analysis and promote critical thinking skills. Where students need to perform calculations or graph data to analyze the results, these steps are also laid out sequentially and in order.

Format and Features
Teacher's Notes

Master Materials List
Lists the chemicals, glassware, and equipment needed to perform the experiment. All amounts have been calculated for a class of 30 students working in pairs. For smaller or larger class sizes or different working group sizes, please adjust the amounts proportionately.

Preparation of Solutions
Calculations and procedures are given for preparing all solutions, based on a class size of 30 students working in pairs. With the exception of particularly hazardous materials, the solution amounts generally include 10% extra to account for spillage and waste. Solution volumes may be rounded to convenient glassware sizes (100 mL, 250 mL, 500 mL, etc.).

Safety Precautions
Repeats the safety precautions given to the students and includes more detailed information relating to safety and handling of chemicals and glassware. Refers to Material Safety Data Sheets that should be available for all chemicals used in the laboratory.

Disposal
Refers to the current *Flinn Scientific Catalog/Reference Manual* for general guidelines and specific procedures governing the disposal of laboratory waste. Because we recommend that teachers review local regulations before beginning any disposal procedure, the information given in this section is for general reference purposes only. However, if a disposal step is included as part of the experimental procedure itself, then the specific solutions needed for disposal are described in this section.

Lab Hints
This section reveals common sources of student errors and misconceptions and where students are likely to need help. Identifies the recommended length of time needed to perform each experiment, suggests alternative chemicals and equipment that may be used, and reminds teachers about new techniques (filtration, pipeting, etc.) that should be reviewed prior to lab.

Teaching Tips
This section puts the experiment in perspective so that teachers can judge in more detail how and where a particular experiment will fit into their curriculum. Identifies the working assumptions about what students need to know in order to perform the experiment and answer the questions. Highlights historical background and applications-oriented information that may be of interest to students.

Sample Data
Complete, actual sample data obtained by performing the experiment exactly as written is included for each experiment. Student data will vary.

Answers to All Questions
Representative or typical answers to all questions. Includes sample calculations and graphs for all data analysis questions. Information of special interest to teachers only in this section is identified by the heading "Note to the teacher." Student answers will vary.

Look for these icons in the *Experiment Summaries and Concepts* section and in the *Teacher's Notes* of individual experiments to identify inquiry-, microscale-, and technology-based experiments, respectively.

Experiment Summaries and Concepts

Experiment

Introduction to Reaction Rates—The "Blue Bottle" Reaction

How fast will a chemical reaction occur? Too slow, and it may not be practical. Too fast, and it may explode! Studying reaction rates helps chemists make a variety of products, from antibiotics to fertilizers, safely and economically. The purpose of this experiment is to show how reaction rates can be measured and to identify conditions that affect the rate of the "blue bottle" reaction of dextrose with methylene blue.

Temperature and Reaction Rates—An Inquiry-Based Approach

How can the rate of a chemical reaction be measured? What effect does changing the temperature have on the rate of a chemical reaction? In this inquiry-based experiment, students must design and carry out a procedure to analyze the rate of reaction of magnesium with hydrochloric acid at different temperatures. What is the independent variable? What is the dependent variable? What variables must be controlled? There's more to a well-planned experiment than just test tubes and chemicals!

The Order of Reaction—Effect of Concentration

Reactions involving iodine and starch are called iodine clock reactions—the blue color appears suddenly, like an alarm clock ringing. When will the clock "ring?" The answer depends on the concentration of reactants! The purpose of this microscale experiment is to determine the order of reaction for an iodine clock reaction by measuring the rate of the reaction starting with different concentrations of a reactant.

Kinetics of Dye Fading—Technology and Graphical Analysis

The color change of phenolphthalein in base is a familiar reaction. But did you know that the color will eventually fade in excess base? In this technology-based experiment, students analyze the kinetics of this "fading" reaction using colorimetry to measure the intensity of the red color as a function of time. Graphing the results reveals how the rate of the fading reaction depends on the concentration of the dye.

Determining a Rate Law—A "Sulfur Clock" Reaction

The rate of a chemical reaction may depend on the concentration of one or more reactants or it may be independent of the concentration of a given reactant. Exactly how the rate depends on reactant concentrations is expressed in an equation called the rate law. The purpose of this advanced-level experiment is to determine the rate law for the acid-catalyzed decomposition of sodium thiosulfate to give elemental sulfur.

Concepts

- Kinetics
- Reaction rate
- Collision theory
- Oxidation–reduction

- Reaction rate
- Temperature
- Kinetic theory
- Collision theory

- Reaction rate
- Rate law
- Order of reaction
- Iodine clock reaction

- Kinetics
- Reaction rate
- Order of reaction
- Colorimetry

- Kinetics
- Rate law
- Order of reaction
- Concentration

Experiment Summaries and Concepts

Demonstration

Iodine Clock Reaction—Effect of Concentration, Temperature, and a Catalyst

Mix a series of two colorless solutions and measure the time until they suddenly change from colorless to deep blue in quick succession. Use this popular iodine clock demonstration to examine the effects of concentration, temperature, and a catalyst on the rate of the reaction. The results make a clear and convincing case for the collision theory of reaction rates.

Now You See It, Now You Don't—Oscillating Chemical Reaction

Surprise your students with this demonstration of an oscillating chemical reaction. Add some white solids to a colorless solution and it quickly changes to orange. Less than a minute later, however, it's back to colorless. The color will continue to oscillate between colorless and orange every 30 seconds for half an hour! What causes the unusual behavior? Take advantage of this surprising reaction to help your students visualize abstract concepts related to the reaction mechanism and reaction intermediates.

Sudsy Kinetics—An "Old Foamey" Demonstration

It's sudsy, it's fun—it's chemical kinetics! Everyone loves the classic "Old Foamey" reaction of hydrogen peroxide. Now, you can also use this exciting demonstration to teach your students about chemical kinetics. How does changing the concentration of a reactant change the reaction rate? What does a catalyst do? Where does a reaction intermediate come from? Find the answers in the cascading foam of Sudsy Kinetics!

The Pink Catalyst—Chemical Demonstration

What does it take to oxidize a simple organic compound? Will mighty hydrogen peroxide do the trick? No—it takes a little pink catalyst! Even strong and mighty chemical reactions sometimes need a little boost from a catalyst to get them going. Learn the tricks of the catalyst trade by studying the color changes of the pink catalyst. You'll see the catalyst as it gets swept up in the reaction pathway, change into something completely different, and then reappear again at the end as if nothing had happened.

The Floating Catalyst—An Enzyme Reaction Demonstration

Almost all chemical reactions that take place in living organisms are catalyzed by enzymes—nature's catalysts. A typical enzyme may make a chemical reaction occur about one million times faster than it would in the absence of a catalyst. Catalase, which breaks down hydrogen peroxide in plant and animal cells, is one of the most active known enzymes. Use this demonstration to show students that what they learn in chemistry lab applies to chemical reactions in living tissue, not just in test tubes.

Concepts

- Reaction rate
- Catalyst
- Collision theory
- Iodine clock reaction

- Reaction mechanism
- Reaction intermediate
- Catalyst
- Oscillating chemical reaction

- Kinetics
- Decomposition reaction
- Reaction intermediate
- Catalyst

- Kinetics
- Catalyst
- Reaction mechanism

- Catalyst
- Enzyme
- Reaction rate
- Concentration

Teacher Notes

Introduction to Reaction Rates
The "Blue Bottle" Reaction

Introduction

How fast will a chemical reaction occur? If a reaction is too slow, it may not be practical. If the reaction is too fast, it may explode. Measuring and controlling reaction rates makes it possible for chemists and engineers to make a variety of products, everything from antibiotics to fertilizers, in a safe and economical manner. The purpose of this experiment is to investigate how the rate of a reaction can be measured and how reaction conditions affect reaction rates.

Concepts

- Kinetics
- Reaction rate
- Collision theory
- Oxidation–reduction

Background

Kinetics is the study of the rates of chemical reactions. As reactants are transformed into products in a chemical reaction, the amount of reactants will decrease and the amount of products will increase. The rate of the reaction can be determined by measuring the concentration of reactants or products as a function of time. In some cases, it is possible to use a simple visual clue to determine a reaction rate. Thus, if one of the reactants is colored but the products are colorless, the rate of the reaction can be followed by measuring the time it takes for the color to disappear. The average rate of the reaction is then calculated by dividing the molar concentration (M) of the colored reactant by the time needed for the color to disappear. Depending on how fast the reaction occurs, the rate would be reported in units of either M/sec or M/min.

Reactions involving the organic dye methylene blue provide a convenient example to study reaction rates. Methylene blue (abbreviated MB) exists in two forms, a reduced form and an oxidized form, that have different colors. The reduced form of methylene blue (MB_{rd}) is colorless, while the oxidized form (MB_{ox}) is blue. The reduced form is easily converted to the oxidized form by mixing it with oxygen in the air (Reaction 1). The oxidized form, in turn, can be converted back to the reduced form by treatment with a reducing agent, such as dextrose, a reducing sugar.

$$MB_{rd} + O_2 \rightarrow MB_{ox} \qquad \textit{Reaction 1}$$
Colorless \qquad *Blue*

In this experiment, we will study the rate of reaction of the blue, oxidized form MB_{ox} with dextrose and potassium hydroxide to give the colorless, reduced form MB_{rd} (Reaction 2). If the initial concentration of MB_{ox} in solution is known, the rate of the reaction can be determined by measuring the time needed for the blue color to disappear.

$$MB_{ox} + \text{dextrose} + KOH \rightarrow MB_{rd} \qquad \textit{Reaction 2}$$
Blue \qquad *Colorless*

The "blue bottle" oxidation reaction of dextrose with methylene blue is a classic demonstration. Oxidation of dextrose may also be carried out with a variety of redox indicators other than methylene blue to give different color changes. See the Teaching Tips *section for a discussion of the mechanism of this reaction and some popular variations.*

Introduction to Reaction Rates – Page 2

Experiment Overview

The purpose of this experiment is to investigate how changing the temperature of the reactants or how changing the concentration of potassium hydroxide will affect the rate of reaction of methylene blue. The basic process is always the same—when a colorless solution containing MB_{rd} is shaken, it turns blue (Reaction 1). The time needed for the solution to turn colorless (Reaction 2) will be measured and will then be used to determine the average rate of reaction.

Pre-Lab Questions

1. Define the terms *oxidation* and *reduction*. *Note:* Consult your textbook, if necessary, for definitions and examples.

2. In the part of this experiment in which methylene blue changes from blue to colorless, is it being oxidized or reduced? What reactant is causing this change? Is this reactant acting as an oxidizing agent or a reducing agent?

3. *Collision theory* offers a simple explanation for how reactions occur—reacting molecules must first collide. In order for colliding molecules to be converted into products, they must collide with enough energy and with a suitable orientation to break existing bonds (in the reactants) and form new bonds (in the products). Any factor that changes either the total number of collisions or the average energy of the colliding molecules should affect the reaction rate.

    (a) Using collision theory, predict how increasing the temperature should affect the rate of a chemical reaction. State the prediction in the form of an if/then hypothesis and give a reason for your hypothesis.

    (b) Using collision theory, predict how increasing the concentration of a reactant should affect the rate of a chemical reaction. State the prediction in the form of an if/then hypothesis and give a reason for your hypothesis.

Materials

"Blue bottle" solution for Part A, 10 mL*
Dextrose solution, $C_6H_{12}O_6$, 0.1 M, 12 mL
Methylene blue solution, 0.1%, 1 mL
Potassium hydroxide solution, KOH, 3.0 M, 6 mL
Water, distilled or deionized
Wash bottle
Labeling or marking pen
Metric ruler
Stopwatch or clock (watch) with second hand

Beakers, 100- or 150-mL, 4†
Beral-type pipets, thin-stem, 4
Graduated cylinder, 10-mL
Hot plate or warm water
Ice or cold water
Test tubes, medium, 3
Test tube rack
Thermometer
Stoppers to fit test tubes, 3

*The blue bottle solution contains dextrose, potassium hydroxide, and methylene blue.
†Several groups may share beakers to make water baths at different temperatures.

Teacher Notes

Safety Precautions

Potassium hydroxide solution is a corrosive liquid; it is particularly dangerous to eyes and may blister and burn skin. Avoid contact with eyes and skin and clean up all spills immediately. Methylene blue is slightly toxic by ingestion. Wear chemical splash goggles and chemical-resistant gloves and apron. The dextrose (sugar) solution will attract ants. Rinse off all work areas with water and wash hands thoroughly with soap and water before leaving the laboratory.

Procedure

Part A. Effect of Temperature

1. Obtain four 100- or 150-mL beakers and make water baths at approximately the following temperatures: 10 °C, 20 °C, 30 °C, and 40 °C. In order to obtain convenient reaction times, avoid temperatures above 40 °C or below 10 °C.

2. Obtain four, thin-stem pipets and place a mark 2 cm from the bottom on each pipet bulb.

3. Fill each pipet bulb to the 2-cm mark with the "blue bottle" solution. Tie a knot in the stem of each pipet to seal it. (See Figure 1.)

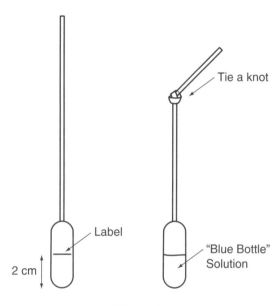

Figure 1.

4. Place one pipet into each of the four water baths from step 1. Let the pipets stand in the bath for 3–5 minutes. Record the temperature of each water bath in the data table.

5. Remove the pipet from the 20 °C water bath, *start timing,* then quickly shake the pipet five times and immediately return it to the water bath.

6. Stop timing when the blue color fades completely and the solution turns colorless. Record the elapsed time in seconds in the data table.

7. Repeat steps 5 and 6 with the other three pipets. Record all time and temperature readings in the data table. *Note:* Try to shake the pipets the same way each time. After shaking, return the pipets to their respective water baths.

8. Dispose of the pipets as directed by your instructor.

Part B. Effect of Concentration

9. Obtain three medium test tubes and stoppers and label them #1–3.

10. Using a graduated cylinder, add 3.0 mL of dextrose solution to each of the three labeled test tubes.

Hot and cold running water should be suitable for preparing water baths in the 10–40 °C temperature range. Have students try to keep the temperature of the baths constant within ± 1 °C by adding more hot or cold water, as needed.

Introduction to Reaction Rates – *Page 4*

11. Add one drop of methylene blue solution to each test tube.

12. Measure 1.0 mL of 3.0 M potassium hydroxide solution into a clean graduated cylinder, then add 2.0 mL of distilled water to get a final volume of 3.0 mL.

13. Pour the contents of the graduated cylinder into test tube #1. Stopper the test tube and shake gently to mix the solutions.

14. Measure 2.0 mL of 3.0 M potassium hydroxide solution into a clean graduated cylinder, then add 1.0 mL of distilled water to get a final volume of 3.0 mL.

15. Pour the contents of the graduated cylinder into test tube #2. Stopper the test tube and shake gently to mix the solutions.

16. Measure 3.0 mL of 3.0 M potassium hydroxide solution into a clean graduated cylinder.

17. Pour the contents of the graduated cylinder into test tube #3. Stopper the test tube and shake gently to mix the solutions.

18. Allow the test tubes to sit undisturbed at room temperature until the blue color fades. *Note:* This may take a few minutes.

19. Check the temperature of the solutions to be sure they are all about the same temperature. Record the temperature in the data table.

20. *With your finger firmly on the stopper,* shake test tube #1 vigorously five times and immediately start timing.

21. Stop timing when the blue color fades completely and the solution turns colorless. Record the elapsed time in seconds in the data table.

22. Repeat steps 20 and 21 using test tube #2 and then again using test tube #3.

23. Dispose of the contents of the test tubes as directed by your instructor.

Teacher Notes

The potassium hydroxide solutions in Part B are very slippery. Students who get some base on their skin may be fooled by the slippery feel and may not realize that the base is actually burning their skin. Wear gloves!

Teacher Notes

Name: _____

Class/Lab Period: _____

Introduction to Reaction Rates

Data Table

Part A. Effect of Temperature				
Temperature, °C				
Reaction Time, sec				

Part B. Effect of Concentration			
Test Tube	1	2	3
Reaction Time, sec			
Temperature, °C			

Post-Lab Questions

1. How did the reaction time change as the temperature was changed in Part A?

2. How is the rate of a reaction related to the time of reaction?

3. What effect does temperature have on the rate of the "blue bottle" reaction?

4. According to a general "rule of thumb" for chemical reactions, the rate of a reaction will roughly double for every 10 °C increase in temperature. Do the kinetics of the "blue bottle" reaction fit this general rule?

Introduction to Reaction Rates – *Page 6*

5. On a separate sheet of paper, make a graph of the results in Part A by plotting the reaction time in seconds on the y-axis versus the temperature in kelvins on the x-axis.

6. Using the graph, estimate how long it would take for the reaction to occur at 275 K and at 325 K. Discuss ways the graph could be improved to give better estimates.

7. Use the "dilution" equation ($M_1V_1 = M_2V_2$) to calculate the concentration of potassium hydroxide in each test tube #1–3 in Part B.

   M_1 = concentration of KOH before mixing V_1 = volume of KOH before mixing
   M_2 = concentration of KOH after mixing V_2 = volume of KOH after mixing

   Sample calculation for test tube #1:

   $$M_2 = \frac{(3.0 \text{ M})(1.0 \text{ mL})}{(6.0 \text{ mL})} = 0.50 \text{ M}$$

8. The concentration of methylene blue in Part B is approximately 2.0×10^{-4} M. Divide the concentration of methylene blue by the reaction time in seconds to calculate the average rate of the reaction in units of M/sec for each test tube #1–3.

9. Does the rate of the "blue bottle" reaction depend on the concentration of potassium hydroxide? Discuss in general terms the effect of reactant concentration on the rate of a chemical reaction.

10. How much did the rate of the reaction change when the concentration of KOH was doubled (test tubes #1 and 2) or tripled (test tubes #1 and 3)?

Teacher Notes

Teacher Notes

Teacher's Notes
Introduction to Reaction Rates

Master Materials List *(for a class of 30 students working in pairs)*

"Blue bottle" solution for Part A, 150 mL*	Beakers, 100- or 150-mL, 20†
Dextrose solution, $C_6H_{12}O_6$, 0.1 M, 250 mL§	Beral-type pipets, thin-stem, 60
Methylene blue solution, $C_{16}H_{18}N_3SCl$, 0.1%, 15 mL§	Graduated cylinders, 10-mL, 15
Potassium hydroxide solution, KOH, 3.0 M, 100 mL§	Hot plates, 3–5, or warm water
Water, distilled or deionized	Ice or cold water
Wash bottles, 15	Test tubes, 15 × 125 mm, 45
Labeling or marking pens, 15	Test tube racks, 15
Stopwatches, 15, or clock with "sweep" second hand	Thermometers, 15
Metric rulers, 15	Stoppers to fit test tubes, 45

*The blue bottle solution contains dextrose, potassium hydroxide, and methylene blue. See the *Preparation of Solutions* section.

†Several groups may share beakers to make water baths at different temperatures.

§For use in Part B.

Preparation of Solutions *(for a class of 30 students working in pairs)*

"Blue Bottle" Solution: Prepare a dextrose solution by dissolving 9 g of anhydrous dextrose (glucose) in about 500 mL of distilled or deionized water. Prepare a potassium hydroxide solution by dissolving 16 g of KOH in about 500 mL of water. Just before class, mix 100 mL of the dextrose solution with 100 mL of the potassium hydroxide solution and add 10 drops of 0.1% methylene blue. *Note:* The blue bottle solution should be prepared fresh at the beginning of each class period. The dextrose and potassium hydroxide solutions may be prepared ahead of time.

Dextrose, 0.1 M: Add 9.0 g of dextrose (glucose) to about 250 mL of distilled or deionized water. Stir to dissolve, then dilute to 500 mL.

Methylene Blue Solution, 0.1%: Add 0.1 g of methylene blue (solid) to 100 mL of distilled or deionized water.

Potassium Hydroxide, 3.0 M: Using an ice bath, cool about 100 mL of distilled or deionized water in a flask. Carefully add 42.1 g of potassium hydroxide pellets and stir to dissolve. Remove the solution from the ice bath and allow it to return to room temperature, then dilute to 250 mL with water.

Safety Precautions

Potassium hydroxide solution is a corrosive liquid and is toxic by ingestion; it is particularly dangerous to eyes and may blister and burn skin. Avoid contact with eyes and skin and clean up all spills immediately. Keep citric acid on hand to neutralize any spills. Methylene blue is slightly toxic by ingestion. Wear chemical splash goggles and chemical-resistant gloves and apron. The dextrose (sugar) solution will attract ants. Rinse off all work areas

---

The advisory board members contribute not only their experiments and demonstrations for these books, but also their lab hints, honed by years of experience. One adviser shared this valuable tip—scavenge garage sales for used coffee urns. They are a great source of hot water for preparing hot water baths without the need for expensive hot plates.

Teacher's Notes

with water and wash hands thoroughly with soap and water before leaving the laboratory. Please consult current Material Safety Data Sheets for additional safety, handling, and disposal information.

Disposal

Consult your current *Flinn Scientific Catalog/Reference Manual* for general guidelines and specific procedures governing the disposal of laboratory waste. The waste solutions from Parts A and B may be flushed down the drain with excess water according to Flinn Suggested Disposal Method #26b.

Lab Hints

- The laboratory work for this experiment can reasonably be completed in one 50-minute class period. The *Pre-Lab Questions* may be assigned separately as preparation for lab, or they may be used as the basis of a cooperative class discussion. If time is short, consider doing Part B as a demonstration.

- Many students will think that the blue–colorless and colorless–blue reactions are the reverse of each other. This is not the case. There are two separate reactions going on—oxidation of the colorless MB_{rd} form to the blue MB_{ox} form by reaction with oxygen, and reduction of the blue MB_{ox} back to the colorless MB_{rd} by reaction with dextrose.

- Reaction of dextrose with methylene blue in the presence of base results in oxidation of the sugar molecule. The aldehyde or hemiacetal functional group in dextrose is oxidized to a carboxylic acid derivative (gluconic acid or gluconolactone). Oxidation of dextrose in this reaction represents an application of the concept of "reducing sugars" that students may be familiar with from prior biology classes. Dextrose is called a reducing sugar because it acts as a reducing agent in reactions with Cu^{2+} or Ag^+ ions (recall the Benedict's test and Tollen's test from carbohydrate chemistry). See the *Supplementary Information* section for the mechanism of oxidation of dextrose.

- The reaction times depend on the number of times the pipet is shaken. Convenient reaction times are obtained if the pipet is shaken about five times. Shaking the pipets 10 or more times gives longer reaction times. The calculated reaction rates are probably not accurate, therefore, in terms of the actual concentration of methylene blue that undergoes reaction. The calculations are used mainly to illustrate how reaction times and reaction rates are related. Although individual rates may not be accurate, the trends in reaction rate as a function of temperature and concentration are reproducible.

- It may be helpful to review beforehand the idea that when the rate of reaction *increases*, the reaction time *decreases*. Using car travel as an analogy usually clarifies the relationship quite effectively.

Teaching Tips

- The "blue bottle" reaction is a classic chemistry demonstration. It is used in general science classes to introduce the roles of observation and hypothesis in the scientific method and in chemistry classes to illustrate oxidation and reduction reactions. It is also a perfect demonstration to talk about the mechanisms or pathways of chemical reactions, which are difficult to study otherwise. Call or write us at Flinn Scientific to obtain a free, complimentary copy of the blue bottle demonstration.

Dextrose, also called glucose, is a monosaccharide, or a simple sugar. It is the main carbohydrate in the human body and also the principal fuel for metabolism.

Teacher's Notes

Teacher Notes

- Other redox indicators may be used instead of methylene blue in this reaction. Indigo carmine is green in its oxidized form, yellow in its reduced form. It gives a green–red–yellow color transition with dextrose. (See the "Stop-n-Go Light—Demonstration Kit," Flinn Catalog No. AP2083.) Resazurin undergoes a reversible red–colorless reaction in the presence of dextrose. (See the "Vanishing Valentine Demonstration Kit," Flinn Catalog No. AP5929.)

Answers to Pre-Lab Questions *(Student answers will vary.)*

1. Define the terms *oxidation* and *reduction*.

   Oxidation and reduction reactions result from the transfer of electrons from one substance to another. Oxidation refers to the process of losing electrons, reduction to the process of gaining electrons.

2. In the part of this experiment in which methylene blue changes from blue to colorless, is it being oxidized or reduced? What reactant is causing this change? Is this reactant acting as an oxidizing agent or a reducing agent?

   Methylene blue is reduced when it changes from blue to colorless. Dextrose is the reactant that causes this change—it is acting as a reducing agent.

3. *Collision theory* offers a simple explanation for how reactions occur—reacting molecules must first collide. In order for colliding molecules to be converted into products, they must collide with enough energy and with a suitable orientation to break existing bonds (in the reactants) and form new bonds (in the products). Any factor that changes either the total number of collisions or the average energy of the colliding molecules should affect the reaction rate.

   (a) Using collision theory, predict how increasing the temperature should affect the rate of a chemical reaction. State the prediction in the form of an if/then hypothesis and give a reason for your hypothesis.

   If the temperature of a reaction increases, then the rate of the reaction should also increase. This hypothesis is based on the idea that increasing the temperature increases the average speed of molecules, which should in turn increase both the number of collisions and, more importantly, the average energy of the collisions.

   (b) Using collision theory, predict how increasing the concentration of a reactant should affect the rate of a chemical reaction. State the prediction in the form of an if/then hypothesis and give a reason for your hypothesis.

   If the concentration of reactants increases, then the rate of the reaction should also increase. This hypothesis is based on the idea that increasing the number of molecules present in solution should increase the rate of collisions between molecules.

Teacher's Notes

Sample Data

Student data will vary.

Data Table

Part A. Effect of Temperature				
Temperature, °C	9 °C	21 °C	32 °C	39 °C
Reaction Time, sec	485 sec	117 sec	37 sec	10 sec

Part B. Effect of Concentration			
Test Tube	1	2	3
Reaction Time, sec	205 sec	106 sec	75 sec
Temperature, °C	21 °C	21 °C	21 °C

Answers to Post-Lab Questions *(Student answers will vary.)*

1. How did the reaction time change as the temperature was changed in Part A?

    Using the 21 °C reaction as a "control" for comparison, the reaction time increased when the pipet was placed in a colder water bath (9 °C), decreased when the pipet was placed in warmer water baths (32 °C or 39 °C).

2. How is the rate of a reaction related to the time of reaction?

    The rate of a reaction is inversely related to the time needed for the reaction to occur. The faster the rate of a reaction, the less time that is required for reactants to be converted to products.

3. What effect does temperature have on the rate of the "blue bottle" reaction?

    The rate of the "blue bottle reaction" appears to be very sensitive to temperature. The reaction rate increased dramatically when the temperature was increased by only 10 °C, and also decreased substantially when the temperature was decreased by 10 °C.

4. According to a general "rule of thumb" for chemical reactions, the rate of a reaction will roughly double for every 10 °C increase in temperature. Do the kinetics of the "blue bottle" reaction fit this general rule?

    The kinetics of the "blue bottle" reaction do not seem to fit this general rule. Every 10 °C temperature rise reduced the reaction time by a factor of 3–4. The rate more than tripled! **Note to teachers:** *The rule is general enough that student may interpret the threefold increase in reaction rate as "roughly double."*

Teacher's Notes

Teacher Notes

5. On a separate piece of paper, make a graph of the results in Part A by plotting the reaction time in seconds on the y-axis versus the temperature in kelvins on the x-axis.

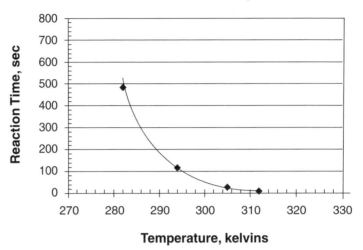

Reaction Time vs. Temperature

6. Using the graph, estimate how long it would take for the reaction to occur at 275 K and at 325 K. Discuss ways the graph could be improved to give better estimates.

   A typical student graph is shown above. Notice that it is almost impossible to predict reaction times at either 275 or 325 K using this graph. In order to estimate the reaction time at 275 K, the y-axis scale would need to be extended beyond 1200 sec (20 min). It is amost impossible to estimate a reaction time at 325 K. The reaction would probably occur so fast that it would be impossible to measure. **Note to teachers:** *Given the current popularity of graphing calculators in math and science classes, many of your students will know how to fit a curve to the data and use the tracer function on their calculators to extrapolate the data.*

7. Use the "dilution" equation ($M_1V_1 = M_2V_2$) to calculate the concentration of potassium hydroxide in each test tube #1–3 in Part B.

   M_1 = concentration of KOH before mixing V_1 = volume of KOH before mixing
   M_2 = concentration of KOH after mixing V_2 = volume of KOH after mixing

   Sample calculation for test tube #1:

   $$M_2 = \frac{(3.0 \text{ M})(1.0 \text{ mL})}{(6.0 \text{ mL})} = 0.50 \text{ M}$$

   For test tube #2: M_2 = (3.0 M)(2.0 mL)/(6.0 mL) = 1.0 M

   For test tube #3: M_2 = (3.0 M)(3.0 mL)/(6.0 mL) = 1.5 M

8. The concentration of methylene blue in Part B is approximately 2.0×10^{-4} M. Divide the concentration of methylene blue by the reaction time in seconds to calculate the average rate of the reaction in units of M/sec for each test tube #1–3.

$$\text{Rate \#1} = \frac{2.0 \times 10^{-4} \text{ M}}{205 \text{ sec}} = 9.8 \times 10^{-7} \text{ M/sec}$$

$$\text{Rate \#2} = \frac{2.0 \times 10^{-4} \text{ M}}{106 \text{ sec}} = 1.9 \times 10^{-6} \text{ M/sec}$$

$$\text{Rate \#3} = \frac{2.0 \times 10^{-4} \text{ M}}{75 \text{ sec}} = 2.7 \times 10^{-6} \text{ M/sec}$$

9. Does the rate of the "blue bottle" reaction depend on the concentration of potassium hydroxide? Discuss in general terms the effect of reactant concentration on the rate of a chemical reaction.

    Yes, the rate of the "blue bottle" reaction depends on the concentration of potassium hydroxide. In general, the rate of a chemical reaction increases when the concentrations of reactants increase.

10. How much did the rate of the reaction change when the concentration of KOH was doubled (test tubes #1 and 2) or tripled (test tubes #1 and 3)?

    The rate increased by a factor of two when the concentration of KOH was doubled, by a factor of three when the concentration was tripled. **Note to teachers:** *The rate appears to be first order with respect to potassium hydroxide, in agreement with the literature.*

Supplementary Information

Oxidation of dextrose (glucose) in the presence of potassium hydroxide involves an initial acid–base reaction to form the glucoside anion, followed by $2e^-$ oxidation to gluconolactone.

D-Glucose → (KOH) → Glucoside Anion → ($-2e^-$, $-H^+$) → D-Gluconolactone

The $2e^-$ oxidation of glucose is coupled with the $2e^-$ reduction of methylene blue (MB_{ox}).

Methylene Blue Oxidized Form (Blue) → ($+2e^-$, $+2H^+$) → Methylene Blue Reduced Form (Colorless)

The products of oxidation of glucose may also include glucuronic acid and degradation products.

Page 1 – **Temperature and Reaction Rates**

Teacher Notes

Temperature and Reaction Rates
An Inquiry-Based Approach

Introduction

The rate of a chemical reaction describes how fast the reaction occurs. How can the rate of a reaction be measured? What effect does temperature have on the rate of a chemical reaction?

Concepts

- Reaction rate
- Temperature
- Kinetic theory
- Collision theory

Background

The greater the rate of a chemical reaction, the less time is needed for a specific amount of reactants to be converted to products. The rate of a reaction can be determined therefore by observing either the disappearance of reactants or the appearance of products as a function of time. Some of the factors that may affect the rates of chemical reactions include the nature of the reactant, the concentration of the reactant, the reaction temperature, the surface area of a solid reactant, and the presence of a catalyst. In this experiment, the effect of temperature on the rate of a chemical reaction will be investigated.

Experiment Overview

The purpose of this inquiry-based experiment is to design and carry out a procedure to determine the effect of temperature on the rate of reaction of magnesium with hydrochloric acid.

Pre-Lab Questions

1. Write the balanced equation for the reaction of magnesium metal with hydrochloric acid.

2. What visible signs of reaction should be observed as the reaction proceeds? How will you be able to determine when the reaction has ended?

3. What measurements must be made to determine the effect of temperature on the rate of the reaction?

4. The independent variable in an experiment is the variable that is changed by the experimenter, while the dependent variable responds to (depends on) changes in the independent variable. Choose the dependent and independent variables for this experiment.

5. What other variables will affect the reaction times in this experiment? How can these variables be controlled?

6. Read the *Materials* section and the recommended *Safety Precautions*. Write a step-by-step procedure for the experiment, including the specific safety precautions that must be followed.

Temperature and Reaction Rates – Page 2

Materials

Copper wire, 18-gauge, 20-cm length, 2*
Hydrochloric acid, HCl, 1 M, 125 mL
Magnesium ribbon, Mg, 24-cm strip†
Metric ruler
Hot plate
Ice
Beakers, 400-mL, 3
Graduated cylinder, 25- or 50-mL
Stopwatch or timer
Test tubes, medium, 6
Test tube rack
Thermometer

*Build copper wire "cages" to keep the magnesium suspended in the hydrochloric acid and to prevent it from floating. The copper wire will not react with the acid.

†Cut into equal-length pieces for the experiment.

Safety Precautions

Hydrochloric acid is a corrosive liquid. Avoid contact with eyes and skin and clean up all spills immediately. Magnesium metal is a flammable solid. **Do not heat the hydrochloric acid directly on a hot plate!** *Heat the hydrochloric acid in a hot water bath. Keep the temperature of the hydrochloric acid between 0 and 60 °C. Do not react magnesium metal with hydrochloric acid in a closed system—do not stopper or cover the test tubes in which the reaction is taking place. Wear chemical splash goggles and chemical-resistant gloves and apron. Wash hands thoroughly with soap and water before leaving the laboratory.*

Procedure

1. Verify the procedure (see the *Pre-Lab Questions*) with your instructor and review all safety precautions.

2. Carry out the procedure and record all data in a suitable data table.

3. Calculate the average reaction rate for each temperature and graph the data appropriately to determine the mathematical relationship, if any, between the rate of the reaction and the temperature.

4. Write a paragraph describing how temperature affects the rate of a chemical reaction. Include in this paragraph a discussion of the possible errors involved in the experiment and their effect on the results.

5. Answer the following *Post-Lab Questions*.

Post-Lab Questions *(Use a separate sheet of paper to answer the following questions.)*

1. According to the kinetic theory, the average kinetic energy of molecules is proportional to their absolute temperature in kelvins. What is the mathematical relationship between the reaction rate and the temperature in kelvins?

2. The collision theory of reaction rates states that the rate of a reaction depends on the number of collisions between molecules, the average energy of the collisions, and the effectiveness of the collisions. Does the effect of temperature on the reaction rate support the collision theory of reaction rates? Explain.

Teacher Notes

Demonstrate how to build a copper wire "cage" to keep the magnesium in solution.

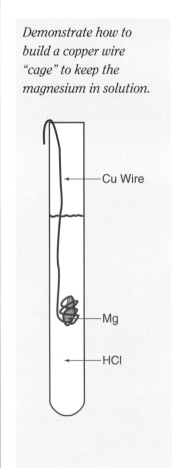

Teacher's Notes
Temperature and Reaction Rates

Master Materials List *(for a class of 30 students working in pairs)*

Copper wire, 18-gauge, 20-cm lengths, 30*
Hydrochloric acid, HCl, 1 M, 2 L
Magnesium ribbon, Mg, 24-cm strips, 15
Metric rulers, 15
Hot plates, 5–6†
Ice†
Stoppers, one-hole (size to fit test tubes), 30§

Beakers, 400-mL, 15–20†
Graduated cylinders, 25- or 50-mL, 15
Stopwatches or timers, 15
Test tubes, 18 × 150 mm, 90
Test tube racks, 15
Thermometers, 15

*Supply copper wire in about 20-cm lengths to the students and show them how to build "cages" to suspend the magnesium ribbon in solution.

†Several groups may share hot-water and ice-water baths.

§One-hole stoppers are optional. They provide a convenient means of hanging the copper wire cages in the test tubes. If one-hole stoppers are not available, the copper wire may simply be hung over the side of the test tube. Do **not** substitute stoppers without holes!

Preparation of Solutions *(for a class of 30 students working in pairs)*

Hydrochloric Acid, 1 M: Carefully add 83 mL of concentrated hydrochloric acid (12 M) to about 500 mL of distilled or deionized water. Stir to mix and allow to cool to room temperature, then dilute to a final volume of 1 L with water. *Note:* Always add acid to water.

Safety Precautions

Hydrochloric acid is a corrosive liquid. Avoid contact with eyes and skin and clean up all spills immediately. Magnesium metal is a flammable solid. **Do not heat the hydrochloric acid directly on a hot plate!** *Heat the hydrochloric acid in a hot water bath. Keep the temperature of the hydrochloric acid between 0 and 60 °C. Do not react magnesium metal with hydrochloric acid in a closed system—do not stopper or cover the test tubes in which the reaction is taking place. Wear chemical splash goggles and chemical-resistant gloves and apron. Wash hands thoroughly with soap and water before leaving the laboratory. Consult current Material Safety Data Sheets for additional safety, handling, and disposal information.*

Disposal

Consult your current *Flinn Scientific Catalog/Reference Manual* for general guidelines and specific procedures governing the disposal of laboratory waste. The waste solutions may be neutralized with sodium hydroxide and rinsed down the drain with excess water according to Flinn Suggested Disposal Method #24b.

Do not make any solid stoppers available during this lab. Size 2, one-hole stoppers (Flinn Catalog No. AP2302) will fit the large test tubes listed on this page.

Teacher's Notes

Lab Hints

- The laboratory work for this experiment can reasonably be completed in one 50-minute lab period. The most important element for success in an inquiry-based activity is student preparation. Sufficient class time should be allotted before lab to think through the measurements that must be made and how the experiment should be conducted. The *Pre-Lab* section contains leading questions to stimulate class discussion.

- To ensure a safe lab environment, it is essential that the teacher provide a mechanism for checking the students' proposed procedures and their understanding of the necessary safety precautions, as recommended in the *Procedure*.

- See the *Supplementary Information* section for a sample procedure and data table. These may be used as an alternative student handout, if desired.

- Encourage student discussion of the optimum number of temperatures and trials for reliable results. The reaction must be carried out at a minimum of three different temperatures—two temperatures will always give a straight line relationship between reaction time and temperature. The reactions are relatively quick—once students feel comfortable with their procedure, additional trials will require only an extra 2–3 minutes. Additional trials may be done either at different temperatures or at the same temperatures to average the results.

- Reaction times are best measured based on the disappearance of the magnesium metal, especially at higher temperatures. Above 50 °C, the evolution of gas bubbles was observed even after all the metal had reacted. This may be due to "outgassing" of dissolved oxygen or hydrogen at higher temperatures. Indeed, it was found that the product mixture obtained from a room temperature run produced bubbles when placed in a 50 °C bath.

- At first glance, the experimental design seems very simple and straightforward. Upon closer examination, however, several factors emerge that have a bearing on the results. In order to isolate the effect of temperature on the reaction rate, it is desirable to carry out the reactions under conditions where the concentration of hydrochloric acid will not change significantly over the course of the reaction. Using 4-cm (0.03-g) strips of magnesium ribbon corresponds to 0.0012 moles of magnesium metal reacting. The amount of hydrochloric acid consumed in the reaction is twice the number of moles of magnesium, or 0.0024 moles. If the volume of 1.0 M hydrochloric acid used is 18 mL, the initial number of moles of HCl present is 0.018 moles, and the amount of HCl consumed is (0.0024/0.018) × 100, or 13% of the total. This is greater than the 5–10% "extent-of-reaction" generally advised for the method of initial rates—the reactant concentration will not be a controlled variable. The surface area of the magnesium metal also changes during the course of the reaction.

Teacher Notes

Our advisers tell us that for an inquiry-based lab, they require the students to submit their proposed procedures the day before the lab. The teachers then check the procedures and return the proofed copies to the students before lab. This ensures that students are prepared and that teachers have time to supervise the actual lab activity, not proof the procedures, during lab time.

Teacher's Notes

Teacher Notes

Teaching Tip

- The inquiry-based approach outlined in this experiment can easily be extended to study other factors that affect the rate of reaction. Similar experiments can be designed to investigate how the concentration of hydrochloric acid, the nature of the metal, and the presence of a catalyst influence the reaction rate. In looking at the effect of HCl concentration, it is probably safest to work down from about 3 M HCl. The relationship between the nature of the metal and the reaction rate can be investigated by comparing the reactions of magnesium, zinc, and aluminum. Finally, the effect of a catalyst on the rate of a reaction can be studied using the reaction of magnesium with water. At room temperature, magnesium reacts very slowly, if at all, with water. Adding sodium chloride significantly increases the reaction rate—chloride ions act as catalysts in reactions at metal surfaces.

Answers to Pre-Lab Questions *(Student answers will vary.)*

1. Write the balanced equation for the reaction of magnesium metal with hydrochloric acid.

   $Mg(s) + 2HCl(aq) \rightarrow MgCl_2(aq) + H_2(g)$

2. What visible signs of reaction should be observed as the reaction proceeds? How will you be able to determine when the reaction has ended?

   The magnesium metal will slowly "disappear" as it reacts with hydrochloric acid to form a soluble magnesium compound, and hydrogen gas bubbles will be observed as the reaction takes place. The reaction is over when all of the metal has reacted and no more gas bubbles are observed. **Note to teachers:** *Students will need additional experience or guidance to determine the best way for measuring the reaction time. Ideally, students will have enough time in the lab to conduct some trial runs and decide whether it is better to look for the metal to disappear or the bubbling to stop. See the* Lab Hints *section for a discussion of this problem.*

3. What measurements must be made to determine the effect of temperature on the rate of the reaction?

   Measure the reaction temperature and the time needed for the magnesium metal to disappear and/or the gas bubbling to stop.

4. The independent variable in an experiment is the variable that is changed by the experimenter, while the dependent variable responds to (depends on) changes in the independent variable. Choose the dependent and independent variables for this experiment.

   Temperature is the independent variable, the time of reaction the dependent variable.

An interesting application of the role of chloride ions as catalysts in reactions at metal surfaces can be found in the design of Flameless Ration Heaters (FRH). A FRH consists of magnesium metal embedded in a polymer matrix with iron (an activator) and sodium chloride (a catalyst). When the water "pouch" in the FRH is broken, the magnesium metal reacts exothermically with the water, producing enough heat to cook a meal.

Temperature and Reaction Rates

Teacher's Notes

5. What other variables will affect the reaction times in this experiment? How can these variables be controlled?

   Other factors that will affect the reaction times include the concentration of the acid, the volume of solution, the amount (mass) of magnesium, and the particle size or surface area of the metal. The experiment should be carried out using only one concentration of hydrochloric acid (1M) and keeping the volume of acid the same in each trial. The magnesium ribbon should be cut into equal-size strips so that the amount of magnesium reacting is the same in each trial. Using magnesium ribbon ensures that the surface area of the metal should be the same in each trial and should not affect the results. Because the magnesium ribbon will be encased in a wire cage to suspend it in solution, it is important that the entire surface of the metal should be exposed to the acid. The magnesium ribbon should be loosely inserted into the wire cages and should not be twisted into a knot.

6. Read the *Materials* section and the recommended *Safety Precautions*. Write a step-by-step procedure for the experiment, including the specific safety precautions that must be followed.

   *See the **Supplementary Information** section for a sample procedure.*

Teacher's Notes

Teacher Notes

Sample Data

Student data will vary.

Data and Results Table

Temperature	2 °C	23 °C	40 °C	53 °C
Reaction Time Trial 1	194 sec	69 sec	58 sec	39 sec
Reaction Time Trial 2	214 sec	76 sec	54 sec	42 sec
Average Reaction Time	204 sec	73 sec	56 sec	41 sec
Reaction Rate, sec^{-1}	4.9×10^{-3}	1.4×10^{-2}	1.8×10^{-2}	2.4×10^{-2}

Sample Graphs

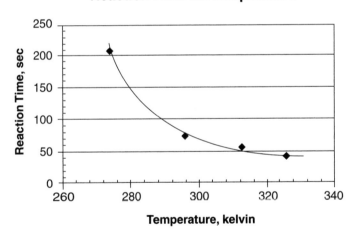

Reaction Time vs. Temperature

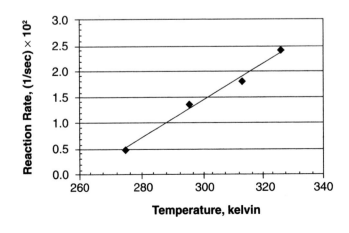

Reaction Rate vs. Temperature

The reaction rates reported in the Data and Results Table are proportional rates obtained by taking the inverse of the reaction time in seconds. Actual reaction rates can be calculated from the number of moles of magnesium that have reacted divided by the reaction time.

Temperature and Reaction Rates

Teacher's Notes

Answers to Post-Lab Questions *(Student answers will vary.)*

1. According to the kinetic theory, the average kinetic energy of molecules is proportional to their absolute temperature in kelvins. What is the mathematical relationship between the reaction rate and the temperature in kelvins?

   The reaction rate appears to be proportional to the temperature, in kelvins, within the temperature range studied. **Note to teachers:** *In general, a linear relationship is not expected.*

2. The collision theory of reaction rates states that the rate of a reaction depends on the number of collisions between molecules, the average energy of the collisions, and the effectiveness of the collisions. Does the effect of temperature on the reaction rate support the collision theory of reaction rates? Explain.

   Increasing the temperature increases the reaction rate, decreasing the temperature decreases the rate, in support of the collision theory of reaction rates. The effect of temperature can be explained in terms of both the number of collisions between molecules and their average energy. As noted in Question #1, increasing the temperature increases the average kinetic energy of molecules—they move faster. As molecules move faster, the rate of collisions between molecules will increase, thus increasing the reaction rate. More importantly, as the average energy of the colliding molecules increases, more of the colliding molecules have sufficient energy to surpass the activation energy barrier and be converted to products.

Teacher's Notes

Teacher Notes

Supplementary Information

Sample Procedure

1. Prepare a hot water bath by filling a 400-mL beaker half-full with water and heating it on a hot plate at a low setting. The temperature of the bath should not exceed 50 °C.

2. Prepare an ice-water bath by filling a 400-mL beaker with water and ice. The temperature of the bath should be between 0 and 5 °C.

3. Using a graduated cylinder, add 18 mL of 1 M hydrochloric acid to each of six large test tubes (approx. 20 × 150 mm). All of the test tubes should be the same size.

4. Place two test tubes in a beaker of room temperature water, two test tubes in the ice water bath, and two test tubes in the hot water bath.

5. Allow the test tubes to sit in their respective water baths for at least 5 minutes to reach thermal equilibrium.

6. Obtain a 24-cm strip of magnesium ribbon.

7. Using scissors, cut the magnesium ribbon into six, 4.0-cm long pieces. *Note:* Be as precise as possible. The reaction time will depend on the amount of magnesium reacting in each trial.

8. Twist and fold one end of the copper wire around a pencil to make a small "cage" into which the magnesium ribbon may be inserted. The other end of the wire must be long enough so that the wire will hang over the side of the test tube and the cage will be below the liquid level marked on the test tube.

9. Measure and record the temperature of the room temperature water bath.

10. Fit one piece of magnesium ribbon loosely through a copper wire cage so the magnesium will be held in place but not wrapped around too tightly.

11. Suspend the copper wire cage and the piece of magnesium in the hydrochloric acid solution in one of the room temperature test tubes and immediately start timing.

12. Measure and record the time until the metal has disappeared and the solution stops bubbling.

13. Repeat steps 10–12 with the other room temperature test tube.

14. Repeat steps 10–13 with each test tube in the ice-water and hot-water baths.

15. Average the reaction times at each temperature.

Which is better—2 runs each at 3 different temperatures or 1 run each at 6 different temperatures? The answer will depend on the temperature stability of the water baths.

Temperature and Reaction Rates

Teacher's Notes

Teacher Notes

*Page 1 – **The Order of Reaction***

Teacher Notes

The Order of Reaction
Effect of Concentration

Introduction

Why does a candle burn more brightly in pure oxygen than it does in air? Oxygen is a reactant in the combustion reaction that takes place when the candle burns. The rate of the reaction, and thus the brightness of the flame, depends on the concentration of oxygen. In this experiment, the effect of concentration on the rate of a chemical reaction will be investigated. The mathematical relationship between the rate and the concentration of a reactant will also be determined.

Concepts

- Reaction rate
- Rate law
- Order of reaction
- Iodine clock reaction

Background

The rate of a reaction describes how fast the reaction occurs—the greater the rate, the less time that is needed for reactants to be converted to products. Reaction rates are therefore inversely proportional to time (rate ∝ 1/time). In general, the rate of a reaction increases as the concentration of reactants increases. The relationship between the rate of a reaction and the concentration of reactants is expressed in a mathematical equation called a *rate law*. For a general reaction of the form

$$A + B \rightarrow C$$

the rate law can be written as

$$\text{Rate} = k[A]^n[B]^m$$

where k is the rate constant, [A] and [B] are the molar concentrations of the reactants, and n and m are exponents that define how the rate depends on the individual reactant concentrations.

The exponents n and m are also referred to as the *order of reaction* with respect to each reactant. In the above example, the reaction is said to be nth order in A and mth order in B. In general, n and m will be positive whole numbers—typical values of n and m are 0, 1, and 2. Note that when $n = 0$, the rate does not depend on the concentration of the reactant. When $n = 1$, the reaction will occur twice as fast when the reactant concentration is doubled, and when $n = 2$, the rate will occur four (2^2) times as fast when the reactant concentration is doubled. The values of the exponents must be determined by experiment—they cannot be predicted simply by looking at the balanced chemical equation.

Consider, for example, the reaction of hydrogen peroxide with iodide ions to give water and iodine (Equation 1). The general rate law for this reaction is given by Equation 2.

$$H_2O_2(aq) + 2I^-(aq) + 2H^+(aq) \rightarrow 2H_2O(l) + I_2(aq) \qquad \textit{Equation 1}$$

$$\text{Rate} = k[H_2O_2]^n[I^-]^m[H^+]^p \qquad \textit{Equation 2}$$

The order of reaction is not limited to positive whole numbers. Half-order reactions ($n = \frac{1}{2}$) are not uncommon, and, in rare cases, the rate may be inversely related to the concentration of a reactant ($n = -1$).

The Order of Reaction

The Order of Reaction – Page 2

The order of reaction with respect to hydrogen peroxide—the value of n—can be determined by measuring the rate of the reaction for several different initial concentrations of hydrogen peroxide. If the concentrations of the other reactants are not changed, the rate will depend only on the concentration of H_2O_2 and the value of n. The general rate law for the reaction will reduce to the form

$$\text{Rate} = k'[H_2O_2]^n \qquad \text{Equation 3}$$

where the constant k' includes the $[I^-]^m$ and $[H^+]^p$ terms.

Experiment Overview

The purpose of this microscale experiment is to investigate the effect of hydrogen peroxide concentration on the rate of its reaction with iodide ions and to determine the reaction order in hydrogen peroxide.

A series of hydrogen peroxide solutions having different concentrations will be prepared. A special microscale "shakedown" technique will be used to mix the hydrogen peroxide solutions with the other reactants—iodide ions, a buffer, and starch. The buffer is used to maintain a constant concentration of H^+ ions. Starch, which forms a characteristic dark blue complex with iodine, is added as an indicator to signal when a specified amount of iodine has been formed. The time needed for the reaction to occur, from the time of mixing to the time when the blue color suddenly appears, will be measured and used to calculate the reaction rate. Reactions involving iodine and starch are called *iodine clock reactions*—the blue color appears suddenly, like an alarm clock ringing.

Pre-Lab Questions

A classic version of the iodine clock reaction involves reaction of iodide ions with persulfate ions (Equation 4).

$$2I^-(aq) + S_2O_8^{2-}(aq) \rightarrow I_2(aq) + 2SO_4^{2-}(aq) \qquad \text{Equation 4}$$

The following rate data was collected by measuring the time required for the appearance of the blue color due to the iodine–starch complex.

Trial	$[I^-]$	$[S_2O_8^{2-}]$	Reaction Time
1	0.040 M	0.040 M	270 sec
2	0.080 M	0.040 M	138 sec
3	0.040 M	0.080 M	142 sec

1. In each trial, the blue color appeared after 0.0020 M iodine (I_2) had been produced. Calculate the *reaction rate* for each trial by dividing the concentration of iodine formed by the reaction time.

2. Compare trials 1 and 2 to determine the order of reaction with respect to iodide ions. How did the concentration of iodide ions change in these two trials, and how did the rate change accordingly? What is the reaction order for iodide?

Teacher Notes

3. Which two trials should be compared to determine the order of reaction with respect to persulfate ions? What is the reaction order for persulfate?

4. Write the rate law for this version of the iodine clock reaction. Could the rate law have been predicted using the coefficients in the balanced chemical equation? Explain.

Materials

Reactants
- Solution A (hydrogen peroxide, H_2O_2, 0.20 M), 2 mL
- Solution B (buffered potassium iodide solution, KI, 0.05 M), 2 mL
- Solution C (starch indicator), 2 mL
- Water, distilled or deionized

Beaker, 50-mL	Microtip pipets, 4
Cassette box (to hold the pipets)	Reaction strips, 12-well, 2
Cotton swabs, 2	Stopwatch, or clock with a second hand
Labels, adhesive (to label pipets), 4	Toothpicks, plastic, 4

Safety Precautions

Dilute hydrogen peroxide is a skin and eye irritant and is slightly toxic by ingestion. Wear chemical splash goggles and chemical-resistant gloves and apron. Wash hands thoroughly with soap and water before leaving the laboratory.

General Microscale Techniques

Proper technique is important when working on the microscale level with very small volumes of liquids. To ensure equal size drops, hold the microtip pipet vertically above the reaction well when dispensing the solutions. Squeeze out a small volume of liquid into a waste beaker before dispensing the counted number of drops into each well. This will remove any air bubbles that may be trapped in the stem of the pipet.

Procedure

Read the entire procedure before beginning the experiment.

1. Obtain four microtip pipets and fold an adhesive label around the stem of each pipet (see Figure 1). Label the pipets A, B, C, and water.

2. Fill each pipet with about 2 mL of the appropriate liquid.

3. Place the pipets in an opened cassette case for storage (Figure 1).

4. Obtain two clean, 12-well reaction strips and arrange them so that the numbers can be read from left to right.

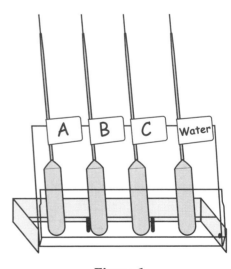

Figure 1

This experiment may work better with students working in groups of three rather than in pairs. The "iodine clock" changes color in quick succession in each series of wells, making it difficult for one person to measure and record the reaction times for all 12 wells.

The Order of Reaction – Page 4

5. Using the following table as a guide, fill each numbered well in the first reaction strip with the appropriate number of drops of Solution A and water. Mix the solution in each well with a new toothpick.

Well	1	2	3	4	5	6	7	8	9	10	11	12
Drops Solution A	4	4	4	3	3	3	2	2	2	1	1	1
Drops of Water	0	0	0	1	1	1	2	2	2	3	3	3

6. To each numbered well in the second reaction strip, add 4 drops of Solution B and 1 drop of Solution C. Mix the solution in each well with a toothpick.

7. Turn the second reaction strip upside down and place it on top of the first strip so that the numbered wells are lined up on top of each other (see Figure 2). *Note:* It seems strange at first, but the liquid will not flow out of the wells until the reaction strips are "snapped" downward in step 8. Surface tension prevents the liquid from flowing out of the wells.

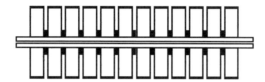

Figure 2

8. Holding the aligned plates firmly together "mouth-to-mouth," as shown above, shake them downward once vigorously with a sharp downward motion. This is done by dropping your hands as fast as you can and stopping abruptly. There is no upward motion—this is a "shakedown" technique. Your partner should start timing *immediately* when the strips are snapped.

9. In the data table, record the time when the solution in each cell turns blue.

10. When all of the cells have turned blue, rinse the contents of the well strips with warm water. Use a cotton swab to dry the inside of each well.

11. Repeat the entire process (steps 4–10) twice.

Teacher Notes

Teacher Notes

Name: _____

Class/Lab Period: _____

The Order of a Reaction

Data and Results Table

Well	1	2	3	4	5	6	7	8	9	10	11	12
[H$_2$O$_2$], M		0.089 M										
Reaction Time Trial 1, sec												
Reaction Time Trial 2, sec												
Reaction Time Trial 3, sec												
Average Reaction Time, sec												
Average Rate, M/sec												

Post-Lab Calculations and Analysis

1. Use the dilution equation ($M_1V_1 = M_2V_2$) to calculate the concentration of hydrogen peroxide in each set of wells #1–3, 4–6, 7–9, and 10–12 after the reaction strips have been "snapped," but before any reaction has occurred. Record the results in the Data and Results Table.

   M_1 = concentration of H$_2$O$_2$ before mixing V_1 = drops of H$_2$O$_2$ before mixing
   M_2 = concentration of H$_2$O$_2$ after mixing V_2 = final number of drops after mixing

   Sample calculation for wells #1–3:

   $$M_2 = \frac{(0.20 \text{ M})(4 \text{ drops})}{(9 \text{ drops})} = 0.089 \text{ M}$$

The concentrations calculated in Question #1 represent the concentration of H$_2$O$_2$ after the two reaction strips have been snapped together, but before any reaction has occurred. The final number of drops is 9 drops in each well.

2. Average the reaction time in seconds for all nine runs at each hydrogen peroxide concentration (three wells per trial, three trials). Record the results in the Data and Results Table.

3. The blue color appears when the amount of iodine that has been produced is 2.8×10^{-4} M. Divide the molar concentration of iodine formed by the average reaction time in seconds to calculate the average rate for each hydrogen peroxide concentration. Enter the results in the table.

$$\text{Average Rate} = \frac{\Delta[I_2]}{\Delta \text{time}} = \frac{2.8 \times 10^{-4} \text{ M}}{\text{average reaction time (sec)}}$$

4. Compare the concentrations of H_2O_2 and the reaction rates for different series of wells to determine the order of reaction in hydrogen peroxide. What is the order of the reaction n in hydrogen peroxide? Explain your reasoning.

5. Briefly describe how the experiment could be modified to determine the order of the reaction in iodide ions.

Teacher Notes

Teacher's Notes
The Order of Reaction

Master Materials List *(for a class of 30 students working in pairs)*

Reactants

Solution A (hydrogen peroxide, H_2O_2, 0.20 M), 30 mL

Solution B (buffered potassium iodide solution, KI, 0.050 M), 30 mL

Solution C (starch indicator), 30 mL

Water, distilled or deionized

Beakers, 50-mL, 15	Microtip pipets, 60
Cassette boxes, 15	Reaction strips, 12-well, 30
Cotton swabs, 30	Stopwatches, 15
Labels, adhesive, 60	Toothpicks, plastic, 60

Preparation of Solutions *(for a class of 30 students working in pairs)*

Solution A (Hydrogen Peroxide, 0.20 M): Using a graduated cylinder, add 23 mL of 3% hydrogen peroxide solution to 50 mL of distilled or deionized water. Stir to mix, then dilute to 100 mL with water.

Solution B (Buffered Potassium Iodide, 0.050 M): Add 1.74 g of potassium iodide, 1.4 g of sodium acetate, and 3.0 mL of 6.0 M acetic acid to about 100 mL of distilled or deionized water. Stir to dissolve, then dilute to 200 mL with water. The final solution is about 0.050 M in potassium iodide and 0.090 M in both sodium acetate and acetic acid. An equimolar mixture of sodium acetate and acetic acid forms a buffer with a pH value of about 4.7.

Solution C (Starch and Sodium Thiosulfate): Boil 100 mL of distilled or deionized water and spray in laundry starch until a faint bluish translucence is observed. Cool the solution to room temperature and add 0.124 g of sodium thiosulfate pentahydrate ($Na_2S_2O_3 \cdot 5H_2O$). *Note:* The concentration of sodium thiosulfate in this stock solution is 0.0050 M. The final concentration of sodium thiosulfate in each reaction well after mixing is 5.6×10^{-4} M. To avoid confusing the students, the thiosulfate content is not disclosed in the student notes. See the *Lab Hints* section for a discussion of the role of sodium thiosulfate in the kinetics of iodine clock reactions.

Safety Precautions

Dilute hydrogen peroxide is a skin and eye irritant and is slightly toxic by ingestion. Wear chemical splash goggles and chemical-resistant gloves and apron. Wash hands thoroughly with soap and water before leaving the laboratory. Please consult current Material Safety Data Sheets for additional safety, handling, and disposal information.

Disposal

Consult your current *Flinn Scientific Catalog/Reference Manual* for general guidelines and specific procedures governing the disposal of laboratory waste. The waste solutions may be flushed down the drain with excess water according to Flinn Suggested Disposal Method #26b.

The KI solution is buffered to maintain a constant concentration of H^+ ions during the reaction. The reaction of H_2O_2 with I^- to give I_2 and H_2O requires acid.

Teacher's Notes

Lab Hints

- The use of microscale techniques in this experiment has many advantages—setup is easy, different concentrations can be tested simultaneously, reaction times are short, and multiple trials can be performed within a typical 50-minute laboratory period. There are, however, some disadvantages to the use of microscale techniques in kinetics, and these disadvantages should be accounted for in the instructional process. With such small volumes, drop size and the absence of air bubbles are absolutely critical. Consider using trial 1 as a practice run to allow students to become comfortable with the procedure. The instructor should demonstrate the "shakedown" technique for the students before they begin the lab.

- Many versions of the iodine clock reaction have been adapted for use in kinetics experiments. Two well-known examples are $I^-/S_2O_8^{2-}$ (see the *PreLab Questions*) and I^-/BrO_3^-. In all cases, the reactions use a limiting quantity of sodium thiosulfate to determine when a threshold concentration of iodine has been produced. (See the *Preparation of Solutions* section). Starch is added as an indicator to detect the formation of iodine above this threshold level. The experiments are usually designed so that the thiosulfate concentration is about 1/10th of the initial iodide concentration. Thiosulfate ions react with any iodine initially formed in the reaction, according to Equation 5. The rate of this second reaction is very rapid compared to the initial slow reaction to form iodine. As a result, iodine is consumed as fast as it is formed. As soon as the limiting quantity of thiosulfate ions has reacted, iodine begins to accumulate. The presence of iodine is detected by the sudden appearance of the dark blue color due to the starch–iodine complex. Using a limiting quantity of thiosulfate ions ensures that the measured rates approximate "initial rates" rather than average rates. Initial rates are preferred for rate law studies because the actual rate decreases over the course of the reaction as the concentration of reactants decreases.

$$2S_2O_3^{2-}(aq) + I_2(aq) \rightarrow S_4O_6^{2-}(aq) + 2I^-(aq) \qquad \text{Equation 5}$$

- In this experiment, the concentration of thiosulfate ions in each well is 5.6×10^{-4} M after mixing. According to Equation 5, the mole ratio for the disappearance of iodine is one mole of iodine per two moles of thiosulfate ions. Multiplying the concentration of thiosulfate ions by this mole ratio gives the amount of iodine that has been produced before the blue color appears— 2.8×10^{-4} M.

- The rationale for the use of thiosulfate ions in iodine clock reactions usually confuses students. Adding the details of the thiosulfate reaction may be frustrating for students who are struggling to understand the kinetics concepts only to find themselves "thrown off the trail" by this second reaction. The details of this aspect of the experimental design have therefore been omitted in the student sections of this write-up. The only practical effect of this omission is that the students must be told what concentration of iodine has been produced when the blue color suddenly appears in the reaction wells.

Teaching Tip

- See the *Supplementary Information* section for optional graphing exercises that can be used to analyze the data and determine the reaction order.

See "The Iodine Clock Reaction" in the Demonstrations section of this book for an alternative version of the iodine clock reaction. The demonstration uses potassium iodate and sodium metabisulfite and does not require sodium thiosulfate. The "clock" color appears when all of the bisulfite ions have been used up.

Teacher's Notes

Teacher Notes

Answers to Pre-Lab Questions *(Student answers will vary.)*

A classic version of the iodine clock reaction involves reaction of iodide ions with persulfate ions (Equation 4).

$$2I^-(aq) + S_2O_8^{2-}(aq) \rightarrow I_2(aq) + 2SO_4^{2-}(aq) \qquad \text{Equation 4}$$

The following rate data was collected by measuring the time required for the appearance of the blue color due to the iodine-starch complex.

Trial	$[I^-]$	$[S_2O_8^{2-}]$	Reaction Time
1	0.04 M	0.040 M	270 sec
2	0.08 M	0.040 M	138 sec
3	0.04 M	0.080 M	142 sec

1. In each trial, the blue color appeared after 0.0020 M iodine (I_2) had been produced. Calculate the *reaction rate* for each trial by dividing the concentration of iodine formed by the reaction time.

    Trial 1: Rate = 0.0020 M/270 sec = 7.4 × 10⁻⁶ M/sec
    Trial 2: Rate = 0.0020 M/138 sec = 1.5 × 10⁻⁵ M/sec
    Trial 3: Rate = 0.0020 M/142 sec = 1.4 × 10⁻⁵ M/sec

2. Compare trials 1 and 2 to determine the order of reaction with respect to iodide ions. How did the concentration of iodide ions change in these two trials, and how did the rate change accordingly? What is the reaction order for iodide?

    In Trials 1 and 2, the concentration of persulfate ions was held constant, while the concentration of iodide ions was doubled. The rate increased by a factor of two when $[I^-]$ was doubled. The reaction is first order in iodide.

3. Which two trials should be compared to determine the order of reaction with respect to persulfate ions? What is the reaction order for persulfate?

    Comparing the rates of Trials 1 and 3 will show how the rate of the reaction depends on the concentration of persulfate ions. In Trials 1 and 3, the concentration of iodide ions was held constant while the concentration of persulfate ions was doubled. The rate increased by a factor of two when $[S_2O_8^{2-}]$ was doubled. The reaction is first order in persulfate.

4. Write the rate law for this version of the iodine clock reaction. Could the rate law have been predicted using the coefficients in the balanced chemical equation? Explain.

    Rate = $k[I^-][S_2O_8^{2-}]$
    The rate law cannot be predicted simply by looking at the balanced chemical equation—the exponents are not the same as the coefficients in the balanced equation.

Teacher's Notes

Sample Data

Student data will vary.

Data and Results Table

Well	1	2	3	4	5	6	7	8	9	10	11	12
$[H_2O_2]$, M	0.089 M			0.067 M			0.044 M			0.022 M		
Reaction Time Trial 1, sec	15	15	19	25	25	25	40	33	36	79	79	82
Reaction Time Trial 2, sec	16	20	21	29	28	34	44	42	39	86	80	81
Reaction Time Trial 3, sec	17	19	20	28	33	27	36	39	42	76	80	82
Average Reaction Time, sec	18 sec			28 sec			39 sec			81 sec		
Average Rate, M/sec	1.6×10^{-5} M/sec			1.0×10^{-5} M/sec			7.2×10^{-6} M/sec			3.5×10^{-6} M/sec		

Answers to Post-Lab Calculations and Analysis *(Student answers will vary.)*

1. Use the dilution equation ($M_1V_1 = M_2V_2$) to calculate the concentration of hydrogen peroxide in each set of wells #1–3, 4–6, 7–9, and 10–12 after the reaction strips have been "snapped," but before any reaction has occurred. Record the results in the Data and Results Table.

   M_1 = concentration of H_2O_2 before mixing
   M_2 = concentration of H_2O_2 after mixing
   V_1 = drops of H_2O_2 before mixing
   V_2 = final number of drops after mixing

   Sample calculation for wells #1–3:

   $$M_2 = \frac{(0.20 \text{ M})(4 \text{ drops})}{(9 \text{ drops})} = 0.089 \text{ M}$$

   For wells #4–6: $M_2 = (0.20 \text{ M})(3 \text{ drops})/(9 \text{ drops}) = 0.067 \text{ M}$

   For wells #7–9: $M_2 = (0.20 \text{ M})(2 \text{ drops})/(9 \text{ drops}) = 0.044 \text{ M}$

   For wells #10–12: $M_2 = (0.20 \text{ M})(1 \text{ drop})/(9 \text{ drops}) = 0.022 \text{ M}$

2. Average the reaction time in seconds for all nine runs at each hydrogen peroxide concentration (three wells per trial, three trials). Record the results in the Data and Results Table.

   See the Sample Data and Results Table.

Flinn ChemTopic™ Labs — Kinetics

Teacher's Notes

3. The blue color appears when the amount of iodine that has been produced is 2.8×10^{-4} M. Divide the molar concentration of iodine formed by the average reaction time in seconds to calculate the average rate for each hydrogen peroxide concentration. Enter the results in the table.

   Sample calculation for wells #1–3:

   Rate = 2.8×10^{-4} M/18 sec = 1.6×10^{-5} M/sec

   See the Sample Data and Results Table for the results of the other calculations.

4. Compare the concentrations of H_2O_2 and the reaction rates for different series of wells to determine the order of reaction in hydrogen peroxide. What is the order of the reaction n in hydrogen peroxide? Explain your reasoning.

   The concentration of hydrogen peroxide in wells #1–3 is two times greater than that in wells #7–9, and the rate is 2.2 times faster. The concentration of hydrogen peroxide in wells #7–9 is two times greater than that in wells #10–12, and the rate is 2.1 times faster. This is pretty close to double the concentration, double the rate, or first order in hydrogen peroxide. The concentration of hydrogen peroxide in wells #4–6 is three times greater than that in wells #10–12, and the rate is 2.9 times faster. This is pretty close to triple the concentration, triple the rate, again suggesting first order in hydrogen peroxide.

5. Briefly describe how the experiment could be modified to determine the order of the reaction in iodide ions.

   In order to determine the reaction order with respect to iodide ions, the experiment could be modified to vary the concentration of iodide ions while holding the hydrogen peroxide concentration the same. This could be done by diluting Solution B (buffered potassium iodide) in the first part of the procedure, then mixing with a constant number of drops of Solutions A and C in the second part of the procedure. ***Note to teachers:*** *Solution B should be diluted with a buffer solution, not with water.*

Supplementary Information

The data in this experiment can also be analyzed graphically. Consider the "reduced" rate law equation for the reaction of hydrogen peroxide and iodide ions (Equation 3 in the *Background* section):

$$\text{Rate} = k'[H_2O_2]^n \qquad \text{Equation 3}$$

The rate of the reaction is inversely proportional to the reaction time in seconds, while the hydrogen peroxide concentration is directly proportional to the number of drops of hydrogen peroxide solution used in wells #1–3, 4–6, 7–9, and 10–12. The rate law can therefore be rewritten as shown in Equation 4.

$$1/\text{time} = \text{constant} \times (\text{drops})^n \qquad \text{Equation 4}$$

The constant in Equation 4 incorporates a multitude of proportionality constants involved in the conversion of the reaction time to the rate of reaction and in the conversion of drops of Solution A to the concentration of hydrogen peroxide. Taking the inverse of both sides of

Teacher's Notes

Equation 4 suggests that a plot of time versus 1/(drops)^n should be a straight line. Comparing the following graphs reveals that the best straight line is obtained by plotting time in seconds versus 1/drops.

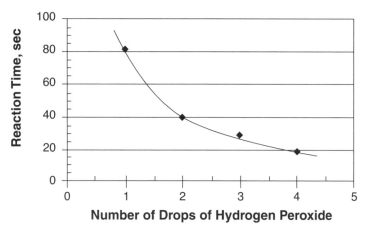

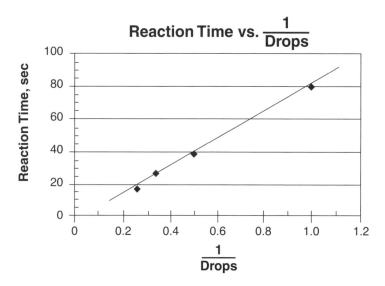

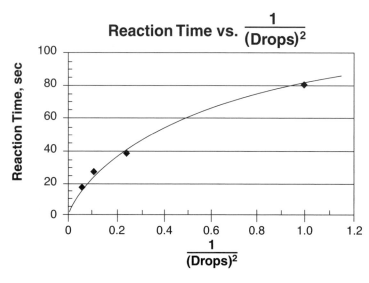

Teacher Notes

Kinetics of Dye Fading
Technology and Graphical Analysis

Introduction

Phenolphthalein is a dye that is used as an acid–base indicator. It is colorless in acidic or neutral solutions and turns bright red-violet as the solution becomes basic. In strongly basic solutions the red color slowly fades and the solution again becomes colorless. The kinetics of this "fading" reaction can be analyzed by measuring the intensity of the red color and graphing the results.

Concepts

- Kinetics
- Reaction rate
- Order of reaction
- Colorimetry

Background

Phenolphthalein is a large organic molecule. In solutions where the pH <8, it has the structure shown in Figure 1, which is colorless. As the solution becomes basic and the pH increases, the phenolphthalein molecule (abbreviated H_2P) loses two hydrogen ions to form the red-violet dianion (abbreviated P^{2-}) shown in Figure 2.

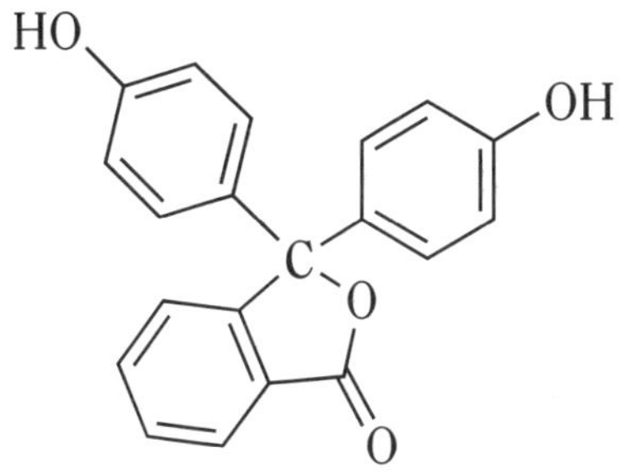

Figure 1. H_2P is colorless.

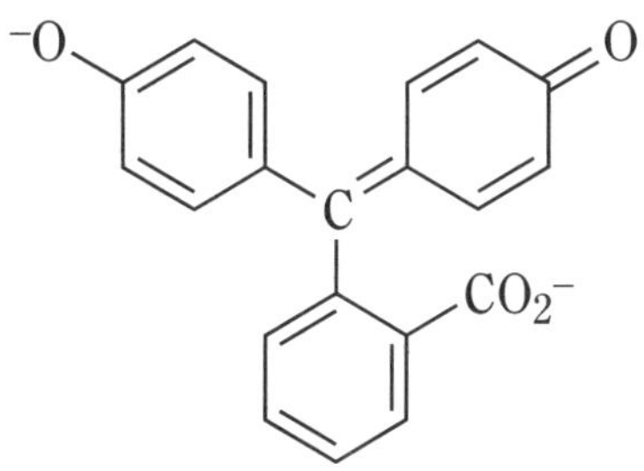

Figure 2. P^{2-} is red.

The colorless-to-red transition of H_2P to P^{2-} (Equation 1) is very rapid and the red color develops instantly when the pH reaches the indicated range. Gradually, however, if the concentration of hydroxide ions remains high, the red P^{2-} dianion will combine with hydroxide ions to form a third species, POH^{3-} (Equation 2), which is also colorless. The rate of this second reaction is much slower than the first and depends on the concentration of phenolphthalein and hydroxide ions.

$$H_2P \xrightarrow{\text{fast}} P^{2-} + 2H^+ \qquad \qquad \textit{Equation 1}$$
Colorless   Red

$$P^{2-} + OH^- \xrightarrow{\text{slow}} POH^{3-} \qquad \qquad \textit{Equation 2}$$
Red   Colorless

The kinetics of the "fading" reaction can be followed by measuring the concentration of P^{2-} versus time and graphing the results. Figure 3 illustrates how the concentration of a reactant decreases with time over the course of a reaction. Notice that the graph of concentration versus time is a curved line, not a straight line. The curve levels off as it approaches the x-axis. This means that the reaction slows down as the reactant concentration decreases.

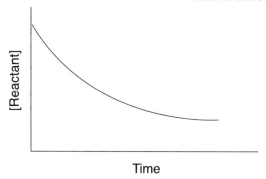

Figure 3.

Exactly how much the rate decreases as the reactant concentration decreases depends on the rate law for the reaction. In the case of the reaction of P^{2-} with OH^- ions, the rate law has the general form

$$\text{Rate} = k[P^{2-}]^n[OH^-]^m \qquad \textit{Equation 3}$$

The exponents n and m are defined as the order of reaction for each reactant and k is the rate constant for the reaction at a particular temperature. The values of the exponents n and m must be determined by experiment. If the reaction is carried out under conditions where the concentration of OH^- does not change—by using a large excess of hydroxide ions—then the rate law will reduce to the form

$$\text{Rate} = k'[P^{2-}]^n \qquad \textit{Equation 4}$$

where k' is a new "pseudo" rate constant incorporating both the "true" rate constant k and the experimentally constant $[OH^-]^m$ term.

Mathematical treatment of the equations for the reaction rate and the rate law predicts the following outcomes:

- If the fading reaction is first order in $[P^{2-}]$ (that is, $n = 1$), a graph of the natural log (ln) of $[P^{2-}]$ versus time will give a straight line. The slope of the line is equal to $-k'$.

- If the fading reaction is second order in $[P^{2-}]$ (that is, $n = 2$), a graph of $1/[P^{2-}]$ versus time will give a straight line. The slope of the line is equal to $-k'$.

Experiment Overview

The purpose of this technology-based experiment is to use colorimetry and graphical analysis to determine how the rate of the phenolphthalein fading reaction depends on the concentration of the dye. A colorimeter is a special instrument that measures the absorbance of light. A known amount of phenolphthalein will be added to a large excess of sodium hydroxide, and the absorbance (Abs) of the red solution will be measured at specific time intervals. Absorbance is directly proportional to concentration, and so a graph of absorbance versus time has the same characteristics as a graph of concentration versus time (Figure 3). Graphing the absorbance data (ln Abs versus time and 1/Abs versus time) should reveal whether the fading reaction is first or second order in phenolphthalein.

Teacher Notes

See the Lab Hints *section for suggestions about how to introduce the basic principles of colorimetry. The color of light absorbed by a substance is complementary to the color of light transmitted or reflected by the substance.*

Page 3 – **Kinetics of Dye Fading**

Teacher Notes

Pre-Lab Questions

Crystal violet (CV) is another indicator dye that combines with hydroxide ions to form a colorless product (Equation 5). Crystal violet was added to 0.10 M NaOH and the solution immediately turned violet. After 10 minutes, the color faded and the solution was almost colorless. The following absorbance measurements were recorded.

$$CV^+ + OH^- \rightarrow CVOH \qquad \qquad \textit{Equation 5}$$
$$\textit{Violet} \qquad \qquad \textit{Colorless}$$

Reaction Time	Absorbance	ln(Abs)	1/Abs
1 min	0.366		
2 min	0.251		
3 min	0.176		
4 min	0.124		
5 min	0.089		
6 min	0.065		
7 min	0.048		
8 min	0.037		
9 min	0.029		
10 min	0.023		

1. Calculate the values of ln(Abs) and 1/Abs for each absorbance measurement to complete the table.

2. Use the following graphs to plot ln(Abs) versus time (Graph 1) and 1/Abs versus time (Graph 2).

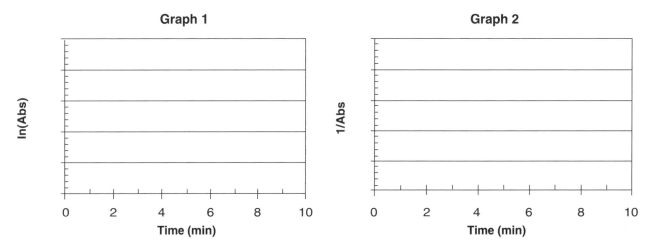

3. Which graph more closely approximates a straight line? Is the reaction of crystal violet with hydroxide ions (Equation 5) first or second order in crystal violet?

Kinetics of Dye Fading – Page 4

Materials

Phenolphthalein solution, 1 drop*
Sodium hydroxide, NaOH, 0.2 M, 5 mL
Thermometer
Tissues or lens paper, lint-free
Wash bottle and distilled water
Cuvette with lid, 1
Colorimeter sensor or spectrophotometer
Computer interface system (LabPro)
Computer or calculator for data collection
Data collection software (LoggerPro)

*The concentration of phenolphthalein is approximately 6×10^{-3} M.

Safety Precautions

Sodium hydroxide is a corrosive liquid. Avoid contact with eyes and skin and clean up all spills immediately. Phenolphthalein is moderately toxic by ingestion. Wear chemical splash goggles and chemical-resistant gloves and apron. Wash hands thoroughly with soap and water before leaving the laboratory.

Procedure

Read the entire procedure before beginning the experiment.

1. Handle the cuvette by its ribbed sides or its top to avoid getting fingerprints on the surface.

2. Connect the interface system to the computer or calculator and plug the colorimeter sensor into the interface.

3. Select *Setup* and *Sensors* from the main screen and choose "Colorimeter."

   Note: Many newer sensors have an automatic calibration feature that automatically calibrates the colorimeter before use. If the sensor has the autocalibration feature, set the wavelength on the colorimeter to 565 nm (green), press the autocalibration key, and proceed to step 9. If the sensor does not have the autocalibration feature, follow steps 4–8 to calibrate the colorimeter for 100% transmission (0 absorbance) with a "blank" cuvette containing only the sodium hydroxide solution.

4. Select *Calibrate* and *Perform Now* from the Experiment menu on the main screen.

5. Obtain a clean and dry cuvette and fill it about ⅔ full with 0.20 M sodium hydroxide. Wipe the cuvette with a lint-free tissue, then place the cuvette in the colorimeter compartment.

6. Set the wavelength knob on the colorimeter to 0%T—the onscreen box should read zero. Press *Keep* when the voltage is steady.

7. Turn the wavelength knob on the colorimeter to 565 nm (green)—the onscreen box should read 100. Press *Keep* when the voltage is steady.

8. Return to the main screen and set up a live readout and data table that will record absorbance as a function of time.

Teacher Notes

The LabPro interface system for collecting laboratory data may be used with a computer, a calculator, or alone. The steps on pp 38–39 represent the computer-interface instructions. The procedure will need to be modified slightly if the LabPro is used as a calculator interface or if a different type of interface system is used.

Teacher Notes

9. Select *Setup* followed by *Data Collection*. Click on *Selected Events* to set the computer for manual sampling.

10. Remove the "blank" cuvette from the colorimeter compartment. Measure and record the initial temperature of the sodium hydroxide solution.

11. Add one drop of phenolphthalein to the cuvette and immediately press *Collect* on the main screen to begin measuring time. This ensures that the absorbance versus time measurements will accurately reflect the time of reaction from the time of mixing.

12. Place the lid on the cuvette and carefully invert the cuvette several times to mix the solution.

13. Place the cuvette in the colorimeter compartment. When one or two minutes have elapsed from the time of mixing, press *Keep* on the main screen to automatically record the absorbance.

14. Continue making absorbance measurements at regular time intervals (at least every two minutes). Press *Keep* on the main screen to automatically record the absorbance at each desired time.

15. When 16 minutes have elapsed from the time of mixing, press *Stop* on the main screen to end the data collection process.

16. If possible, save the data on the computer or calculator and obtain a printout of the absorbance versus time data table and graph. Otherwise, record the absorbance and time measurements in the data table.

17. Remove the cuvette from the colorimeter compartment. Measure and record the final temperature of the dye solution.

18. Rinse the cuvette several times with distilled water and allow it to air dry.

19. *(Optional)* Calculations and graphical analysis may be done on the computer or calculator using either the data collection software that accompanies the technology interface system or a conventional spreadsheet program such as Excel. See the *Post-Lab Questions* section.

It is important to wash and rinse the plastic cuvettes immediately after the timed run is over. Sodium hydroxide will react with or be absorbed by the plastic if left in the cuvette too long.

Name: _____

Class/Lab Period: _____

Kinetics of Dye Fading

Data Table

Initial Temperature		Final Temperature	
Time	Absorbance*	Ln(Abs)*	1/Abs*

*Computer-generated tables and graphs may be substituted for the data table and *Post-Lab Questions* 1–3.

Post-Lab Questions *(Use a separate sheet of paper to answer the following questions.)*

1. Plot or obtain a graph of absorbance versus time. Does the "rate of fading" of phenolphthalein depend on the concentration of the dye? Explain.

2. Calculate the values of ln(Abs) and 1/Abs for each absorbance measurement and enter the results in the table. *Note:* This may be done directly with the data saved in the technology program or separately using a calculator or spreadsheet program.

3. Plot or obtain graphs of both ln(Abs) versus time and of 1/Abs versus time. (See the *Background* section and the *Pre-Lab Questions*.)

4. Which graph more closely approximates a straight line?

5. Is the reaction of phenolphthalein with hydroxide ions (Equation 2) first or second order in phenolphthalein?

6. Did the temperature of the solution change over the course of the reaction? What effect, if any, would the temperature change have on the results of the experiment?

7. The concentration of sodium hydroxide is assumed to be constant throughout the reaction and is thus included in the "reduced" rate law expression (see Equation 4 in the *Background* section). Is this assumption valid? Prove it.

Teacher Notes

Many teachers prefer to have students draw their own graphs. Graphing the data themselves will help students practice an important skill and may, in fact, make it easier for students to interpret the data and results.

Teacher's Notes
Kinetics of Dye Fading

Master Materials List *(for a class of 30 students working in pairs)*

Phenolphthalein, dilute solution, 2 mL	Cuvettes with lids, 15
Sodium hydroxide, NaOH, 0.2 M, 75 mL	Colorimeter sensors, 15
Thermometers, 15	Computer interface system (LabPro), 15
Tissues or lens paper, lint-free	Computers or calculators for data collection, 15
Wash bottles and distilled water, 15	Data collection software (LoggerPro)

Preparation of Solutions *(for a class of 30 students working in pairs)*

Sodium Hydroxide, 0.20 M: Carefully add 0.80 g of sodium hydroxide to 50 mL of distilled or deionized water and stir to dissolve. Dilute to 100 mL with water.

Phenolphthalein Dilute Solution: Prepare or obtain a standard 1% solution of phenolphthalein in ethyl alcohol (Flinn Catalog No. P0012). Dilute this standard phenolphthalein solution by a factor of five with ethyl alcohol—5 mL of standard 1% solution and 20 mL of ethyl alcohol, for example. *Note:* The final concentration of phenolphthalein is about 6×10^{-3} M.

Safety Precautions

Sodium hydroxide is a corrosive liquid. Avoid contact with eyes and skin and clean up all spills immediately. Phenolphthalein is moderately toxic by ingestion. Wear chemical splash goggles and chemical-resistant gloves and apron. Wash hands thoroughly with soap and water before leaving the laboratory. Please consult current Material Safety Data Sheets for additional safety, handling, and disposal information.

Disposal

Consult your current *Flinn Scientific Catalog/Reference Manual* for general guidelines and specific procedures governing the disposal of laboratory waste. The dye solutions may be flushed down the drain with excess water according to Flinn Suggested Disposal Method #26b.

Lab Hints

- The actual amount of lab time needed for this experiment is about 30–40 minutes. An additional 20–30 minutes of computer time will be required if the calculations and graphing are done using the data collection software that accompanies the technology interface system. Alternatively, the calculations and graphing may be assigned as homework. The difference between the ln(Abs) and 1/Abs graphs is very distinct even in less precise hand-drawn graphs. The ln(Abs) versus time graph gives an excellent straight line fit, while the 1/Abs versus time graph is curved.

- Although this experiment serves a useful role in helping to meet technology goals for the curriculum, it is very challenging conceptually. Graphical analysis of kinetic data to determine whether a reaction is first or second order is usually reserved for honors-level or even advanced placement chemistry courses. The *Background* section contains a summary of the use of graphical analysis in kinetic studies. A more detailed explanation is provided in the *Supplementary Information* section.

The concentration of a 1% phenolphthalein solution is about 0.03 M. If starting with 0.5% phenolphthalein, dilute 10 mL to 25 mL with ethyl alcohol. To prepare a 1% solution of phenolphthalein, dissolve 1 g of phenolphthalein in 100 mL of ethyl alcohol.

Teacher's Notes

- A review of the principles of light absorption and transmission will be necessary if this is the first colorimetry experiment in your lab program. The phenolphthalein solution absorbs light in the 350–600 region. The wavelength of light used in this experiment is 565 nm, corresponding to green light. The beam of light is passed through the sample, and the intensity of the light that is transmitted is measured electronically. The greater the concentration of phenolphthalein in solution, the more green light the solution will absorb. See the "Color and Light Spectrum Demonstrations Kit" (Flinn Catalog No. AP6172) for a large-scale demonstration of the relationship between the color of absorbed and transmitted light.

- Absorbance data may also be collected continuously *via* a timed run, as opposed to manually at selected time intervals. The drawback is that the temperature of the solution may increase more if the light source is on continuously.

- The experiment may also be performed using a conventional spectrophotometer rather than colorimetry to measure absorbance as a function of time.

Teaching Tip

- This experiment can be extended to determine the reaction order with respect to hydroxide ions. Have different student groups collect absorbance data at different hydroxide ion concentrations. For each hydroxide ion concentration, a graph of ln(Abs) versus time should be linear with a slope equal to $-k'$ (see the Background section), where k' incorporates the hydroxide concentration ($k' = k[OH^-]^m$). The classroom data collected by different student groups can be compared to determine the reaction order in hydroxide ion. A graph of k' versus $[OH^-]$ is linear and demonstrates that the reaction is first order with respect to hydroxide ions ($m = 1$). See the *Supplementary Information* section for results and graphical analysis of the effect of hydroxide ion concentration on the rate of fading of phenolphthalein.

Teacher's Notes

Teacher Notes

Answers to Pre-Lab Questions *(Student answers will vary.)*

Crystal violet (CV) is another indicator dye that combines with hydroxide ions to form a colorless product (Equation 5). Crystal violet was added to 0.10 M NaOH and the solution immediately turned violet. After 10 minutes, the color faded and the solution was almost colorless. The following absorbance measurements were recorded.

$$CV^+ + OH^- \rightarrow CVOH$$
Violet   *Colorless*   *Equation 5*

Reaction Time	Absorbance	ln(Abs)	1/Abs
1 min	0.366	–1.01	2.73
2 min	0.251	–1.38	3.98
3 min	0.176	–1.74	5.68
4 min	0.124	–2.09	8.06
5 min	0.089	–2.42	11.2
6 min	0.065	–2.73	15.4
7 min	0.048	–3.04	20.8
8 min	0.037	–3.30	27.0
9 min	0.029	–3.54	34.5
10 min	0.023	–3.77	43.5

1. Calculate the values of ln(Abs) and 1/Abs for each absorbance measurement to complete the table.

   See the table.

2. Use the following graphs to plot ln(Abs) versus time (Graph 1) and 1/Abs versus time (Graph 2).

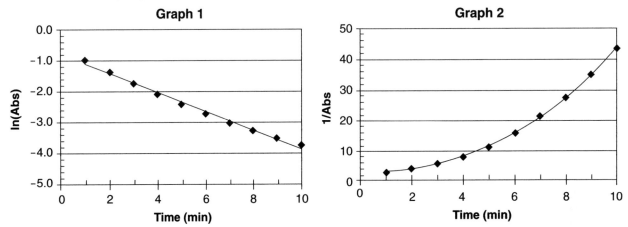

3. Which graph more closely approximates a straight line? Is the reaction of crystal violet with hydroxide ions (Equation 5) first or second order in crystal violet?

   The graph of ln(Abs) versus time gives a straight line. The graph of 1/Abs versus time is curved. The reaction is first order in crystal violet.

Kinetics of Dye Fading

Teacher's Notes

Sample Data

Student data will vary.

Data Table

Initial Temperature	21.5 °C	Final Temperature	22.8 °C
Time	Absorbance*	Ln(Abs)*	1/Abs*
1 min	0.610	−0.49	1.64
2 min	0.545	−0.61	1.84
4 min	0.428	−0.84	2.34
6 min	0.327	−1.12	3.06
8 min	0.247	−1.40	4.05
10 min	0.188	−1.67	5.31
12 min	0.146	−1.93	6.86
14 min	0.116	−2.16	8.66
15 min	0.104	−2.27	9.66

*Computer-generated tables and graphs may be substituted for the data table.

Answers to Post-Lab Questions *(Student answers will vary.)*

1. Plot or obtain a graph of absorbance versus time. Does the "rate of fading" of phenolphthalein depend on the concentration of the dye? Explain.

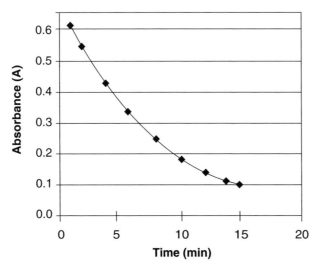

Yes, the "rate of fading" depends on the concentration of the dye. The graph of absorbance vs. time is curved, suggesting that the rate decreases as the concentration of phenolphthalein decreases over the course of the reaction.

The graph of ln(Abs) vs. time is linear with a slope = −.129 (k′ = .129) and correlation coefficient = .999.

Teacher's Notes

Teacher Notes

2. Calculate the values of ln(Abs) and 1/Abs for each absorbance measurement and enter the results in the table. *Note:* This may be done directly with the data saved in the technology program or separately using a calculator or spreadsheet program.

   See the Data Table.

3. Plot or obtain graphs of both ln(Abs) versus time and of 1/Abs versus time. (See the *Background* section and the *Pre-Lab Questions*.)

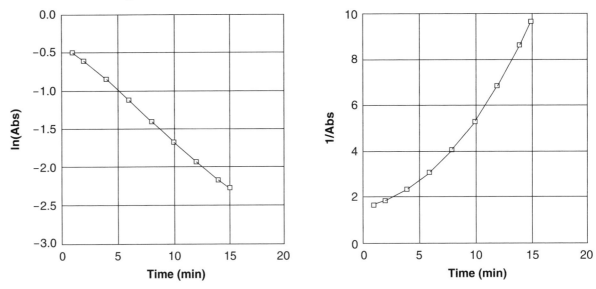

4. Which graph more closely approximates a straight line?

   The points on the ln(Abs) versus time graph all fall on a straight line. The 1/Abs versus time graph is curved.

5. Is the reaction of phenolphthalein with hydroxide ions (Equation 2) first or second order in phenolphthalein?

   The reaction is first order in phenolphthalein.

6. Did the temperature of the solution change over the course of the reaction? What effect, if any, would the temperature change have on the results of the experiment?

   The temperature of the solution increased from 21.5 to 22.8 °C over the course of the reaction. In general, the rate of a reaction increases as the temperature increases. However, the temperature increase in this case is fairly small and does not seem to affect the results. **Note to teachers:** *The temperature effect is probably due to "heating" of the sample by the light source. To reduce the temperature effect, remove the cuvette from the colorimeter compartment between measurements. The temperature change does not seem to be large enough to warrant the increased probability that the solution might spill as it is repeatedly inserted and removed.*

7. The concentration of sodium hydroxide is assumed to be constant throughout the reaction and is thus included in the "reduced" rate law expression (see Equation 4 in the *Background* section). Is this assumption valid? Prove it.

   The initial concentration of phenolphthalein in the cuvette is roughly 1×10^{-4} M— one drop of 0.006 M phenolphthalein is diluted with about 3 mL (60 drops) of sodi-

Teacher's Notes

um hydroxide. The initial concentration of sodium hydroxide is 0.2 M, more than 1000× greater than the phenolphthalein concentration. The amount of sodium hydroxide consumed during the reaction is insignificant—the assumption is valid.

Supplementary Information

Graphing Exercises

Graphical analysis of first and second order reactions is based on combining the equations for the disappearance of reactant (Equation 6) and the rate law (Equation 7) for a pseudo-first order reaction.

$$\text{Rate} = \frac{-\Delta[A]}{\Delta t} \qquad \text{Equation 6}$$

$$\text{Rate} = k'[A]^m \qquad \text{Equation 7}$$

The resulting equation is

$$\frac{-\Delta[A]}{\Delta t} = k'[A]^m \quad \text{or} \quad \frac{-\Delta[A]}{[A]^m} = k'\Delta t$$

Using calculus, it can be shown that if $m = 1$, the so-called integrated rate equation is $\ln[A] = -k't + \ln[A]_o$. A graph of $\ln[A]$ versus time will be linear with slope $-k'$.

If $m = 2$, the integrated rate equation has the form $1/[A] = k't + 1/[A]_o$. A graph of $1/[A]$ versus time will be linear with slope k'.

Order of Reaction in Hydroxide Ion

The following ln(Abs) vs. time graphs were obtained for different hydroxide ion concentrations.

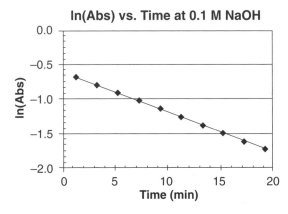
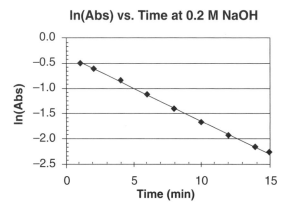

The value of k' was equal to 0.13 for 0.2 M NaOH, 0.061 for 0.1 M NaOH. Substituting these values into the equation for the pseudo-rate constant ($k' = k[OH^-]^m$) gives the following results.

At 0.2 M NaOH　　　　　At 0.1 M NaOH

$$.13 = k' = k[0.2]^m \qquad .061 = k' = k[0.1]^m$$

$$\frac{.13}{.061} = 2 = \left(\frac{0.2}{0.1}\right)^m$$

$$m = 1$$

The fading reaction of phenolphthalein is first order in sodium hydroxide.

Teacher Notes

Determining a Rate Law
A "Sulfur Clock" Reaction

Introduction

The rate of a chemical reaction may depend on the concentration of one or more reactants or it may be independent of the concentration of a given reactant. Exactly how the rate depends on reactant concentrations is expressed in an equation called the rate law. How can the rate law for a reaction be determined?

Concepts

- Kinetics
- Rate law
- Order of reaction
- Concentration

Background

For a general reaction of the form

$$A + B \rightarrow C \qquad \text{Equation 1}$$

the rate law can be written as

$$\text{Rate} = k\,[A]^n[B]^m \qquad \text{Equation 2}$$

where k is the rate constant, [A] and [B] are the molar concentrations of the reactants, and n and m are exponents that define how the rate depends on the individual reactant concentrations. The values of n and m must be determined by experiment—they cannot be determined simply by looking at the balanced chemical equation. The rate constant for a reaction does not depend on the reactant concentrations, but does depend on temperature. The exponents n and m are also referred to as the *order of reaction* with respect to a particular reactant. In the above example, the reaction is said to be nth order in A and mth order in B. In general, n and m will be positive whole numbers—typical values of n and m are 0, 1, 2.

Rate laws for different reactions take on different forms. The reactions shown below and their experimentally determined rate laws demonstrate that the order of reaction cannot be predicted using the coefficients in the balanced chemical equation.

Chemical Equation	Rate Law
$NO_2(g) + CO(g) \rightarrow NO(g) + CO_2(g)$	Rate = $k\,[NO_2]^2$
$2NO(g) + O_2(g) \rightarrow 2NO_2(g)$	Rate = $k\,[NO]^2[O_2]$
$2N_2O_5(g) \rightarrow 4NO_2(g) + O_2(g)$	Rate = $k\,[N_2O_5]$

The reaction order for each reactant in the rate law determines how the rate changes as the concentration of that reactant changes. If a reactant has an order of zero, the reactant does not appear in the rate law and the rate is independent of its concentration. Increasing or decreasing the concentration of a zero-order reactant does not affect the rate of the reaction.

This is an advanced-level experiment. The analysis involved in the experimental design assumes a working knowledge of reaction rates and how they are measured.

Determining a Rate Law – Page 2

When a reaction is first order in a reactant, the reactant appears in the rate law with an exponent of one—the rate is directly proportional to the reactant concentration. If the concentration of a first-order reactant is doubled, the rate will also double. For a reaction that is second order in a reactant, the reactant concentration appears in the rate law with an exponent of two. If the concentration of a second-order reactant is doubled, the rate of the reaction will increase by a factor of four.

The following general procedure may be used to determine the rate law for a reaction. First, the concentration of one reactant is held constant while the concentration of a second reactant is varied, and the reaction time is measured. Then, the first reactant's concentration is varied while the second reactant's concentration is held constant, and again the reaction time is measured. The average rate for each reaction is calculated by taking the inverse of the reaction time. The data is then analyzed to determine the reaction order for each reactant and the rate law.

Consider, for example, the reaction between nitric oxide and oxygen gas.

$$2NO(g) + O_2(g) \rightarrow 2NO_2(g) \qquad \text{Equation 3}$$

$$\text{Rate} = k\,[NO]^n[O_2]^m \qquad \text{Equation 4}$$

The following data was obtained by performing the reaction five times and varying the concentrations of the reactants as indicated. In each case, the reaction time was measured, then inverted to find the average rate.

Trial	[NO]	[O_2]	Reaction Time (sec)	Average Rate (sec^{-1})
1	0.020	0.010	35.7	0.028
2	0.020	0.020	17.5	0.057
3	0.020	0.040	8.8	0.11
4	0.040	0.020	4.4	0.23
5	0.010	0.020	71.4	0.014

In the first three trials, the concentration of NO was constant while the concentration of O_2 was varied. Therefore, any change in the rate in Trials 1–3 is due solely to the change in the concentration of O_2. Comparing the rates in Trials 1 and 2 gives the following results, where the subscripts 1 and 2 refer to Trials 1 and 2, respectively.

$$\frac{\text{Rate}_2}{\text{Rate}_1} = \frac{[O_2]_2^m}{[O_2]_1^m} = \left(\frac{[O_2]_2}{[O_2]_1}\right)^m$$

$$\frac{.057}{.028} = 2.0 = \left(\frac{.02}{.01}\right)^m = 2^m$$

$$m = 1$$

Teacher Notes

The "average rates" shown in the data table are actually "proportional" rates obtained by taking the inverse of the reaction time. Actual rates have the units of concentration divided by time. As long as the reaction times are measured for the appearance of a constant amount of product, as in this experiment, the "inverse time" rates will, in fact, be proportional to the actual rates.

Teacher Notes

In trials 2, 4, and 5 the concentration of O_2 was held constant while the concentration of NO was varied. Therefore, any change in the rate in Trials 2, 4, and 5 is due solely to the change in the NO concentration. In the rate law, the $k[O_2]^m$ terms can be ignored because they do not vary. Comparing the rates in Trials 4 and 5 gives the following results.

$$\frac{\text{Rate}_4}{\text{Rate}_5} = \frac{[NO]_4^n}{[NO]_5^n} = \left(\frac{[NO]_4}{[NO]_5}\right)^n$$

$$\frac{.227}{.014} = 16 = \left(\frac{.04}{.01}\right)^n = 4^n$$

$$n = 2$$

The reaction of NO and O_2 to give NO_2 is first order in O_2, second order in NO. The overall rate equation for the reaction is rate $= k[NO_2]^2[O_2]$.

Experiment Overview

The purpose of this experiment is to determine the rate law for the reaction between hydrochloric acid, HCl, and sodium thiosulfate, $Na_2S_2O_3$.

$$2HCl(aq) + Na_2S_2O_3(aq) \rightarrow S(s) + SO_2(aq) + H_2O(l) + 2NaCl(aq) \quad \textit{Equation 5}$$

In Part A, the HCl concentration will be held constant while the $Na_2S_2O_3$ concentration is varied. In Part B, the HCl concentration will be varied while the $Na_2S_2O_3$ concentration is held constant. Reaction times will be measured by monitoring the appearance of sulfur. As solid sulfur is produced in the reaction, the reaction mixture will become clouded with a yellow precipitate. The reaction time will be measured by noting the time at which it is no longer possible see through the solution.

Determining a Rate Law – Page 4

Pre-Lab Questions

Read the *Background* section and the *Procedure,* then answer the following questions.

1. The following table summarizes the amounts and concentrations of the reactants that will be used in each trial in Parts A and B. Use the dilution equation $M_1V_1 = M_2V_2$ to calculate the concentration M_2 of each reactant in each well after mixing but *before* any reaction occurs. The first one has been worked for you as an example.

Sample		Volume of 1.0 M HCl	Volume of Water	Volume of 0.30 M $Na_2S_2O_3$
Part A	Well 1	2.0 mL	0	3.0 mL
	Well 2	2.0 mL	1.5 mL	1.5 mL
	Well 3	2.0 mL	2.0 mL	1.0 mL
Part B	Well 4	3.0 mL	0	2.0 mL
	Well 5	1.5 mL	1.5 mL	2.0 mL
	Well 6	1.0 mL	2.0 mL	2.0 mL

M_1 = concentration of reactant before mixing

M_2 = concentration of reactant after mixing

V_1 = volume of reactant before mixing

V_2 = volume of reactant after mixing

For well #1: $M_2(HCl) = (1.0 \text{ M})(2.0 \text{ mL})/(5.0 \text{ mL}) = 0.40 \text{ M}$

$M_2(Na_2S_2O_3) = (0.30 \text{ M})(3.0 \text{ mL})/(5.0 \text{ mL}) = 0.18 \text{ M}$

2. Enter the results of the calculations in the data table.

Materials

Hydrochloric acid solution, 1.0 M, HCl, 20 mL
Sodium thiosulfate solution, 0.30 M, $Na_2S_2O_3$, 20 mL
Distilled or deionized water and wash bottle
Beakers, 50-mL, 3
Labeling or marking pen

Cotton swabs or paper towels, 4
Piece of white paper
Reaction plate, six-well
Stopwatch or timer
Syringe, 3-mL

Safety Precautions

Hydrochloric acid solution is moderately toxic by ingestion and inhalation. It is corrosive to eyes and skin. Sodium thiosulfate is a body tissue irritant. The sulfur produced in this reaction has low toxicity, but may be a skin and mucous membrane irritant. The reaction generates aqueous sulfur dioxide, which is a skin and eye irritant. Wear chemical splash goggles and chemical-resistant gloves and apron. Wash hands thoroughly with soap and water before leaving the laboratory.

Teacher Notes

The final volume of solution is 5.0 mL in each well.

Page 5 – **Determining a Rate Law**

Teacher Notes

Procedure

Preparation

1. Label one small beaker "HCl". Add 20 mL of 1.0 M hydrochloric acid to this beaker.

2. Label another small beaker "Na$_2$S$_2$O$_3$". Pour 20 mL of 0.30 M sodium thiosulfate solution into this beaker.

3. Label a third small beaker "water". Pour 30 mL of distilled or deionized water into this beaker.

4. On a piece of white paper draw a black "+" sign about the size of a well in the reaction plate. Verify that the "+" sign can be seen through the plate. Use the same "+" sign for every well reaction in Parts A and B.

Part A. Varying the Concentration of Na$_2$S$_2$O$_3$

Read the entire procedure before beginning the experiment.

5. Fill the 3-mL syringe up to the 2.0-mL mark with distilled water by submerging the syringe in the "water" beaker and drawing water into the syringe until the bottom of the plunger is at the 2.0-mL mark. Make sure there are no air bubbles in the syringe.

6. Now submerge the syringe in the "Na$_2$S$_2$O$_3$" beaker and draw 1.0 mL of the Na$_2$S$_2$O$_3$ solution into the syringe so that the plunger sits at the 3.0-mL mark.

7. Empty the syringe into well #3 of the six-well reaction plate. See Figure 1 below.

8. Repeat steps 5–7 to fill wells #1 and #2. Use the amounts of water and Na$_2$S$_2$O$_3$ solution shown below in Figure 1. Fill the wells in reverse order (well #2 next, then well #1). Filling the syringe with the most dilute mixture first and working up to the most concentrated means that the syringe does not need to be rinsed between fillings.

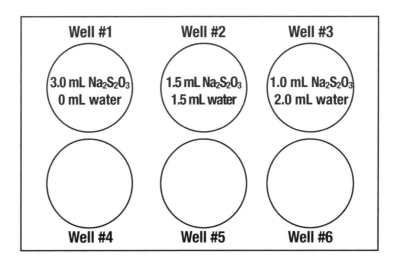

Figure 1.

If more syringes are available, have students label and use two syringes, one for sodium thiosulfate and a second for hydrochloric acid.

Determining a Rate Law

Determining a Rate Law – Page 6

9. Rinse the syringe thoroughly with water. Fill the 3-mL syringe to the 2.0-mL mark with the HCl solution. Prepare to start the timer. Empty the syringe into well #1. Time the reaction with a stopwatch by measuring the time from which the solution was added until the black "+" sign can no longer be seen through the solution. Record the exact time in seconds in the data table.

10. Repeat Step 9 for wells #2 and #3, adding 2 mL of the HCl solution to each well. Carefully time each reaction with a stopwatch by measuring the time from which the solution was added until the black "+" sign can no longer be seen through the solution. Record the exact time in seconds in the data table.

Part B. Varying the Concentration of HCl

11. Fill the 3-mL syringe to the 2.0-mL mark with distilled water by submerging the syringe in the "water" beaker and drawing water into the syringe until the plunger is at the 2.0-mL mark. Make sure that there are no air bubbles in the syringe.

12. Now submerge the syringe in the "HCl" beaker and draw 1.0 mL of HCl solution into the syringe so that the plunger sits at the 3-mL mark.

13. Empty the syringe into the well #6 of the six-well reaction plate. See Figure 2 below.

14. Repeat steps 11–13 to fill wells #5 and #6. Use the amounts of water and HCl solution shown below in Figure 2. Fill the wells in reverse order (well #5 next, then well #4).

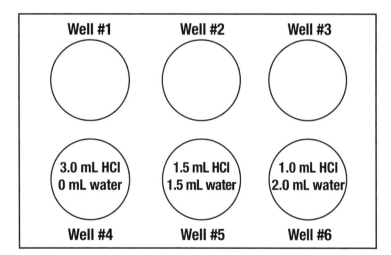

Figure 2.

Page 7 – **Determining a Rate Law**

Teacher Notes

15. Rinse the syringe thoroughly with water. Fill the 3-mL syringe to the 2.0-mL mark with the $Na_2S_2O_3$ solution. Prepare to start the timer. Empty the syringe into well #4. Time the reaction with a stopwatch or timer by measuring the time from which the solution was added until the black "+" sign can no longer be seen through the solution. Record the exact time in seconds in the data table.

16. Repeat Step 15 for wells #5 and #6, adding 2.0 mL of the $Na_2S_2O_3$ solution to each well. Carefully time each reaction with a stopwatch or timer by measuring the time from which the solution was added until the black "+" sign can no longer be seen through the solution. Record the exact time in seconds in the data table.

17. As soon as the reaction times have been measured, quickly empty the six-well reaction plate into the collection container provided by your teacher. Rinse and dry each of the wells with soap and water. Use a cotton swab or a paper towel to thoroughly clean and dry each well.

Determining a Rate Law – Page 8

Name: _____

Class/Lab Period: _____

Determining a Rate Law

Data Table

	Well	[Na$_2$S$_2$O$_3$]*	[HCl]*	Reaction Time
Part A	1			
	2			
	3			
Part B	4			
	5			
	6			

*These are the reactant concentrations in solution immediately after mixing and before any reaction has occurred. See the *Pre-Lab Questions* for calculations.

Post-Lab Questions *(Use a separate sheet of paper to answer the following questions.)*

1. The average rate of reaction is equal to the molar concentration of sulfur produced when the solution becomes cloudy divided by the reaction time.

$$\text{Rate} = \frac{[S]}{\text{Reaction Time}}$$

   If the concentration of sulfur produced in each well is the same at the onset of "cloudiness," then the rate is proportional to 1/time.

$$\text{Rate} \propto \frac{1}{\text{Time}}$$

   Calculate the "proportional" rate in 1/sec for each well.

2. Does the rate depend on the Na$_2$S$_2$O$_3$ concentration? Compare the concentration of Na$_2$S$_2$O$_3$ and the rate in wells #1–3. What is the reaction order for Na$_2$S$_2$O$_3$?

3. Does the rate depend on the HCl concentration? Compare the concentration of HCl and the rate in wells #4–6. What is the reaction order for HCl?

4. Write the rate law for this reaction.

Teacher Notes

Teacher's Notes
Determining a Rate Law

Master Materials List *(for a class of 30 students working in pairs)*

Hydrochloric acid solution, 1.0 M, HCl, 500 mL

Sodium thiosulfate solution, 0.30 M, $Na_2S_2O_3$, 500 mL

Syringes, 3-mL, 15	Reaction plates, six-well, 15
Beakers, 50-mL or other small size, 45	Stopwatches or timers, 15
Labeling or marking pens, 15	Wash bottles, 15
Cotton swabs or paper towels, 60	White background paper, 15

Distilled or deionized water, 1 L

Preparation of Solutions *(for a class of 30 students working in pairs)*

Hydrochloric Acid, 1.0 M: Carefully add 83 mL of concentrated hydrochloric acid (12 M) to about 500 mL of distilled or deionized water. Stir to mix and cool to room temperature, then dilute to a final volume of 1 L. *Note:* Always add acid to water.

Sodium Thiosulfate, 0.30 M: Add 74.5 g of sodium thiosulfate pentahydrate ($Na_2S_2O_3 \cdot 5H_2O$) to about 500 mL of distilled or deionized water. Stir to dissolve, then dilute to 1 L with water.

Safety Precautions

Hydrochloric acid solution is moderately toxic by ingestion and inhalation. It is corrosive to eyes and skin. Sodium thiosulfate is a body tissue irritant. The sulfur produced in this reaction has low toxicity, but may be a skin and mucous membrane irritant. Aqueous sulfur dioxide is generated in this reaction, which is a skin and eye irritant. Wear chemical splash goggles, chemical-resistant gloves, and a chemical-resistant apron. Please consult material safety data sheets for additional safety and handling techniques.

Disposal

Have students empty their six-well reaction plates into one large collection container. Filter the collected solution and dispose of the solid in a landfill according to Flinn Suggested Disposal Method #26a. Neutralize and dispose of the filtrate according to Flinn Suggested Disposal Method #24b. Please consult your current *Flinn Scientific Catalog/Reference Manual* for proper disposal procedures.

Lab Hints

- Make sure that the students empty the well plates and clean them thoroughly using cotton swabs or paper towels after they have finished the experiment. If the plates are allowed to sit with the sulfur precipitate in them, the precipitate will begin to deposit on the bottom of the wells. It will then be difficult to thoroughly clean the wells.

Teacher's Notes

- It is important to add the contents of the second syringe-full of solution with a reasonable amount of force so that good mixing occurs between the solutions in the well. But, if the syringe is emptied with too much force, the solutions will splash out of the desired well. Therefore, urge students to find an emptying rate that ensures good mixing without spillage.

- As the beakers containing the solutions become emptied, it will be harder to fill the syringes without introducing air bubbles into them. Remind students to tilt the beakers so that the liquid can be drawn up into the syringe without air bubbles.

- If six-well reaction plates are not available, beakers, medicine cups, or other small clear collection containers may be used instead. The containers should be small enough to ensure good mixing of the reactants.

- When lower concentrations of sodium thiosulfate are used, the rate law does not appear to be as simple as predicted in this experiment. At lower concentrations, the reaction appears to be closer to 3/2-order in thiosulfate and half-order in hydrochloric acid. The reaction time is more difficult to measure at lower concentrations because the onset of turbidity is more gradual.

- Both the overall reaction equation and the mechanism for the decomposition of sodium thiosulfate are more complex than suggested by this experiment. The reaction is acid-catalyzed, which means that the acid concentration must have some bearing on the rate in terms of producing an equilibrium concentration of $HS_2O_3^-$ ions. The $HS_2O_3^-$ ion is a reactive intermediate, reacting further with additional $S_2O_3^{2-}$ ions to produce polymeric ions containing multiple S atoms. When the chain of S atoms in a polymeric ion becomes long enough, it "closes" in on itself to form a ring of elemental sulfur (S_8).

$$S_2O_3^{2-} + H^+ \rightleftharpoons HS_2O_3^-$$

$$H-S-SO_3^- + nS_2O_3^{2-} \rightarrow H-S-(S)_n-SO_3^- + nSO_3^{2-}$$

$$H-S-S_n-SO_3^- \rightleftharpoons H^+ + {}^-S-S_n-SO_3^-$$

$$^-S-S_8-SO_3^- \rightarrow S_8 + SO_3^{2-}$$

Teaching Tips

- This experiment can easily be performed as a demonstration experiment using small beakers on the overhead projector. The amounts of reagents may be scaled up proportionally to give better visibility, and the reaction times will also increase. Do not use more than about 5 mL of sodium thiosulfate solution per trial—the smell of the sulfur may become irritating.

- Exponents in the rate law cannot be determined from a net balanced chemical equation because the equation represents the sum of several elementary steps in the overall reaction mechanism. Each of these steps will play a role in determining the rate. For each of the elementary steps, however, the rate law can be obtained by making the exponent for each reactant equal to its coefficient in the net balanced chemical equation.

Teacher's Notes

Answers to Pre-Lab Questions

1. The following table summarizes the amounts and concentrations of the reactants that will be used in each well in Parts A and B. Use the dilution equation $M_1V_1 = M_2V_2$ to calculate the concentration M_2 of each reactant in each trial solution after mixing but *before* any reaction occurs. The first one has been worked for you as an example.

Sample		Volume of 1.0 M HCl	Volume of Water	Volume of 0.30 M $Na_2S_2O_3$
Part A	Well 1	2.0 mL	0	3.0 mL
	Well 2	2.0 mL	1.5 mL	1.5 mL
	Well 3	2.0 mL	2.0 mL	1.0 mL
Part B	Well 4	3.0 mL	0	2.0 mL
	Well 5	1.5 mL	1.5 mL	2.0 mL
	Well 6	1.0 mL	2.0 mL	2.0 mL

M_1 = concentration of reactant before mixing

M_2 = concentration of reactant after mixing

V_1 = volume of reactant before mixing

V_2 = volume of reactant after mixing

For Well #1: $M_2(HCl) = (1.0\ M)(2.0\ mL)/(5.0\ mL) = 0.40\ M$

$M_2(Na_2S_2O_3) = (0.30\ M)(3.0\ mL)/(5.0\ mL) = 0.18\ M$

For Well #2: $M_2(HCl) = (1.0\ M)(2.0\ mL)/(5.0\ mL) = 0.40\ M$

$M_2(Na_2S_2O_3) = (0.30\ M)(1.5\ mL)/(5.0\ mL) = 0.090\ M$

For Well #3: $M_2(HCl) = (1.0\ M)(2.0\ mL)/(5.0\ mL) = 0.40\ M$

$M_2(Na_2S_2O_3) = (0.30\ M)(1.0\ mL)/(5.0\ mL) = 0.060\ M$

For Well #4: $M_2(HCl) = (1.0\ M)(3.0\ mL)/(5.0\ mL) = 0.60\ M$

$M_2(Na_2S_2O_3) = (0.30\ M)(2.0\ mL)/(5.0\ mL) = 0.12\ M$

For Well #5: $M_2(HCl) = (1.0\ M)(1.5\ mL)/(5.0\ mL) = 0.30\ M$

$M_2(Na_2S_2O_3) = (0.30\ M)(2.0\ mL)/(5.0\ mL) = 0.12\ M$

For Well #6: $M_2(HCl) = (1.0\ M)(1.0\ mL)/(5.0\ mL) = 0.20\ M$

$M_2(Na_2S_2O_3) = (0.30\ M)(2.0\ mL)/(5.0\ mL) = 0.12\ M$

2. Enter the results of the calculations in the Data Table.

*See the **Sample Data** section.*

Teacher's Notes

Sample Data

Student data will vary.

Data Table

	Well	[Na$_2$S$_2$O$_3$]*	[HCl]*	Reaction Time
	1	0.18 M	0.40 M	17 sec
Part A	2	0.09 M	0.40 M	33 sec
	3	0.06 M	0.40 M	57 sec
	4	0.12 M	0.60 M	19 sec
Part B	5	0.12 M	0.30 M	21 sec
	6	0.12 M	0.20 M	21 sec

*These are the reactant concentrations in solution immediately after mixing and before any reaction has occurred. See the *Pre-Lab Questions* for calulations.

Answers to Post-Lab Questions *(Student answers will vary.)*

1. The average rate of reaction is equal to the molar concentration of sulfur produced when the solution becomes cloudy divided by the reaction time.

$$\text{Rate} = \frac{[S]}{\text{Reaction Time}}$$

If the concentration of sulfur produced in each well is the same at the onset of "cloudiness," then the rate is proportional to 1/time.

$$\text{Rate} \propto \frac{1}{\text{Time}}$$

Calculate the "proportional" rate in 1/sec for each well.

Well #1: Rate = 1/17 sec = .059 sec^{-1}
Well #2: Rate = 1/33 sec = .030 sec^{-1}
Well #3: Rate = 1/57 sec = .018 sec^{-1}
Well #4: Rate = 1/19 sec = .053 sec^{-1}
Well #5: Rate = 1/21 sec = .048 sec^{-1}
Well #6: Rate = 1/21 sec = .048 sec^{-1}

Teacher's Notes

Teacher Notes

2. Does the rate depend on the $Na_2S_2O_3$ concentration? Compare the concentration of $Na_2S_2O_3$ and the rate in wells #1–3. What is the reaction order for $Na_2S_2O_3$?

   The rate increases as the $S_2O_3^{2-}$ concentration increases. Comparing wells 1 and 2 shows that the rate increased by a factor of two when the $S_2O_3^{2-}$ concentration was doubled. Comparing wells 1 and 3 shows that the rate increased by a factor of 3.3 when the $S_2O_3^{2-}$ concentration was tripled. The reaction appears to be first order in $Na_2S_2O_3$ (n = 1).

3. Does the rate depend on the HCl concentration? Compare the concentration of HCl and the rate in wells #4–6. What is the reaction order for HCl?

   The rate does not depend on the HCl concentration. The reaction appears to be zero order in HCl (m = 0).

4. Write the rate law for this reaction.

   Rate = $k\,[S_2O_3^{2-}]$

   Note to teachers: *This is the best estimate based on the results of this experiment. The decomposition reaction is acid-catalyzed—no reaction occurs in the absence of acid. The reaction mixture may also effectively be buffered due to the production of sulfite and bisulfite ions (SO_3^{2-} and HSO_3^{-}). The H^+ concentration would likely appear in the rate "constant," $k' = k_{acid}[H^+] + k_{base}[OH^-]$.*

Teacher's Notes

Teacher Notes

Iodine Clock Reaction
Effect of Concentration, Temperature, and a Catalyst

Introduction

Mix a series of two colorless solutions and measure the time until they suddenly change from colorless to deep blue in quick succession. Use this popular iodine clock demonstration to examine the effects of concentration, temperature, and a catalyst on the rate of a reaction.

Concepts

- Reaction rate
- Collision theory
- Catalyst
- Iodine clock reaction

Materials

Potassium iodate, KIO_3, 0.20 M, 325 mL

Sodium metabisulfite, $Na_2S_2O_5$, 0.20 M, 60 mL*

Starch solution, 180 mL†

Sulfuric acid, H_2SO_4, 0.1 M, 10 mL

Water, distilled or deionized

Balance, centigram

Beakers, 250-mL, 6

Beakers, 400-mL, 6

Graduated cylinder, 10-mL

Graduated cylinder, 50-mL

Graduated cylinder, 100-mL

Hot plate

Ice bath

Thermometer

Timer or stopwatch

Stirring rod

*Sodium metabisulfite solution has a poor shelf life. Prepare within 1–2 weeks of its use.

†Prepare fresh before use. See the *Tips* section.

Safety Precautions

Potassium iodate is an oxidizer. It is moderately toxic by ingestion and a body tissue irritant. Sodium metabisulfite is a skin and body tissue irritant. Sulfuric acid solution is corrosive to eyes, skin and other tissues. Wear chemical splash goggles, chemical-resistant gloves, and a chemical-resistant apron. Please review current Material Safety Data Sheets for additional safety, handling, and disposal information.

Preparation

1. Label six 400-mL beakers as 1A, 2A, 3A, 4A, 5A, and 6A and place them in order on the demonstration table.

2. Using Table 1 as a guide, add the appropriate amounts of 0.2 M KIO_3, distilled water, and 0.1 M of H_2SO_4 to each beaker. Check to make sure each beaker contains 200 mL of solution.

3. Place beaker 4A on a hot plate and warm it to about 45 °C. Place beaker 5A into an ice bath and cool it to about 10 °C.

Iodine Clock Reaction

Demonstrations

Table 1

Beaker	1A	2A	3A	4A	5A	6A
0.2 M KIO_3	50 mL	100 mL	25 mL	50 mL	50 mL	50 mL
Distilled Water	150 mL	100 mL	175 mL	150 mL	150 mL	140 mL
0.1 M H_2SO_4	—	—	—	—	—	10 mL

4. Label six 250-mL beakers as 1B, 2B, 3B, 4B, 5B, and 6B.

5. Pour 10 mL of 0.20 M sodium metabisulfite, 30 mL of the starch solution, and 40 mL of distilled or deionized water into each beaker. Stir each solution.

6. Prepare a data table similar to Table 2 on the chalkboard or on an overhead.

Table 2

Reaction	1	2	3	4	5	6
$[KIO_3]$*	0.04 M	0.07 M	0.02 M	0.04 M	0.04 M	0.04 M
$[Na_2S_2O_5]$*	0.007 M	0.007 M	0.007 M	0.007 M	0.007 M	0.007 M
Temperature	Room Temp	Room Temp	Room Temp	Warm	Cool	Room Temp
Catalyst Added	No	No	No	No	No	Yes
Reaction Time						

*These are the final concentrations of reactants after mixing solutions A and B but *before* any reaction has occurred.

Procedure

1. Pour Solution 1B into Solution 1A. Carefully time the reaction with a stopwatch or timer. Measure the time from when the two solutions are mixed until the appearance of the blue color. Record the reaction time in the data table. *(This is the control reaction—the rates of the other reactions will be compared against this one.)*

Effect of Concentration on the Rate

2. Pour Solution 2B into Solution 2A. Measure and record the time until the appearance of the blue color. Compare the concentration of KIO_3 and the rate of Reaction 2 versus the control. *(The time will be half that of the control—the reaction occurs twice as fast when the concentration is doubled.)*

3. Pour Solution 3B into Solution 3A. Measure and record the time until the appearance of the blue color. Compare the concentration of KIO_3 and the rate of Reaction 2 versus the control. *(The time should be double that of the control—the reaction occurs half as fast when the concentration is halved.)*

Teacher Notes

Many students are initially confused about the relationship between the reaction time and the reaction rate. A reaction that takes twice as long is only half as fast!

Flinn ChemTopic™ Labs — Kinetics

Demonstrations

Teacher Notes

Effect of Temperature on the Rate

4. Carefully remove Beaker 4A from the hot plate and measure the temperature of the solution. Immediately pour Solution 4B into Solution 4A. Measure and record the time until the appearance of the blue color. *(The time should be shorter than that of the control—increasing the temperature increases the rate.)*

5. Remove Beaker 5A from the ice bath and measure the temperature of the solution. Immediately pour Solution 5B into Solution 5A. Measure and record the time until the appearance of the blue color. *(The time should be longer than that of the control—decreasing the temperature decreases the rate.)*

Effect of a Catalyst on the Rate

6. Pour Solution 6B into Solution 6A. Measure and record the time until the appearance of the blue color. What effect does adding a catalyst (sulfuric acid) have on the reaction rate? *(The time should be much shorter than that of the control.)*

Results

Beaker	1	2	3	4	5	6
Reaction Time	6 sec	3 sec	12 sec	4 sec	8 sec	2 sec

Disposal

Please consult your current *Flinn Scientific Catalog/Reference Manual* for general guidelines and specific procedures governing the disposal of laboratory wastes. The final solutions may be disposed of according to Flinn Suggested Disposal Method #26b.

Tips

- Prepare 180 mL of starch solution by making a smooth paste of 3.6 g soluble starch and 20 mL distilled or deionized water. Pour the paste into 160 mL of boiling water while stirring. Cool to room temperature before using. Prepare this solution within 1–2 days.

- The final concentrations of the reactants listed in Table 2 were calculated using the dilution equation, $M_1V_1 = M_2V_2$. The values correspond to the concentrations after mixing solutions A and B but before any reaction has occurred.

Discussion

Many teachers report success in using spray starch to prepare a starch solution. Boil 100 mL of distilled or deionized water, spray in laundry starch until a faint bluish translucence is observed, and allow to cool.

Potassium iodate and sodium metabisulfite react to form iodine. Starch, which forms a characteristic dark-blue colored complex with iodine, is added as an indicator to signal the end of the reaction. The overall reaction occurs in a series of steps, as outlined below.

Step 1: Formation of iodate ions (IO_3^-) and hydrogen sulfite ions (HSO_3^-) in solution.

Solution A: $KIO_3(aq) \rightarrow IO_3^-(aq) + K^+(aq)$

Solution B: $H_2O(l) + Na_2S_2O_5(s) \rightarrow 2HSO_3^-(aq) + 2Na^+(aq)$

Iodine Clock Reaction

Demonstrations

Step 2: Iodate ions react with hydrogen sulfite ions to produce iodide ions (I⁻).

$$IO_3^-(aq) + 3HSO_3^-(aq) \rightarrow I^-(aq) + 3H^+(aq) + 3SO_4^{2-}(aq)$$

Step 3: In the presence of hydrogen ions (H⁺), the iodide ions react with excess iodate ions to produce iodine (I_2). This is the kinetically slow step in the overall reaction.

$$6H^+(aq) + 5I^-(aq) + IO_3^-(aq) \rightarrow 3I_2(aq) + 3H_2O(l)$$

Step 4: Before the iodine can react with the starch to produce a dark-blue colored complex, it immediately reacts with any hydrogen sulfite ions still present to form iodide ions.

$$I_2(aq) + HSO_3^-(aq) + H_2O(l) \rightarrow 2I^-(aq) + SO_4^{2-}(aq) + 3H^+(aq)$$

Step 5: Once all of the hydrogen sulfite ions have reacted, the iodine is then free to react with the starch to form the familiar dark-blue colored complex.

$$I_2(aq) + starch \rightarrow dark\text{-}blue\ colored\ complex$$

The appearance of the dark-blue color in solution indicates that all of the reactants have been used up and the reaction has gone to completion. Therefore, the rate of reaction can be measured by recording the time to the appearance of the dark-blue color.

The effect of concentration, temperature, and a catalyst on the rate of a reaction can be understood within the framework of the collision theory of reaction rates. *Collision theory* offers a simple explanation for how fast a reaction will occur—in order for a reaction to occur, reactant molecules must first collide. Not all collisions, however, will lead to products. In order for colliding molecules to produce products, the collision energy must exceed a certain critical energy level, called the activation energy, for the reaction. If the activation energy is low, almost all colliding molecules will have sufficient energy to overcome the energy barrier for the reaction. These reactions will occur very fast. In contrast, if the activation energy is high, only a small fraction of the colliding molecules will have sufficient energy to overcome the energy barrier for the reaction. These reactions will occur much more slowly.

To increase the rate of a reaction, one of two things must occur: (1) more molecules with sufficient kinetic energy to overcome the activation energy barrier must collide, or (2) the height of the activation energy barrier must be reduced.

Changing the concentration or the temperature of reactants influences the rate of collisions between molecules and the average energy of the collisions, respectively. Increasing the concentration of the reactants increases the rate of collisions between molecules, which in turn increases the total number of effective collisions as well (even though the fraction of collisions that are effective does not change). Increasing the temperature increases the average kinetic energy of the colliding molecules, since kinetic energy is proportional to the absolute temperature in kelvins. Thus, the fraction of molecules with sufficient kinetic energy to overcome the activation energy barrier increases as the temperature increases.

A catalyst increases the rate of a reaction because it decreases the activation energy that is needed for reactants to be transformed to products. In general, a catalyst provides a modified or new pathway for the reaction to occur. The new reaction pathway has a lower activation energy. When the activation energy for the reaction is reduced, the fraction of colliding molecules that have enough energy to overcome this energy barrier increases.

Teacher Notes

See "The Pink Catalyst" on page 72 for an alternative demonstration of what a catalyst does and how it works.

Demonstrations

Teacher Notes

Now You See It — Now You Don't
An Oscillating Chemical Reaction

Introduction

Surprise your students with this demonstration of an oscillating chemical reaction. Add some white solids to a colorless solution and it quickly changes color to orange. Less than a minute later, it's back to colorless. The color will continue to oscillate between colorless and orange every 30 seconds for half an hour. What causes this unusual behavior? Take advantage of the surprising results to help students visualize abstract concepts related to the reaction mechanism.

Concepts

- Reaction mechanism
- Catalyst
- Reaction intermediate
- Oscillating chemical reaction

Materials

Malonic acid, $CH_2(CO_2H)_2$, 4.5 g
Manganous sulfate, $MnSO_4 \cdot H_2O$, 1.3 g
Potassium bromate, $KBrO_3$, 4.0 g
Sulfuric acid solution, H_2SO_4, 1.5 M, 400 mL

Balance, centigram
Beaker, 1-L
Magnetic stirrer and magnetic stir bar
Plastic weighing dishes, 3

Safety Precautions

Sulfuric acid solution is severely corrosive to eyes, skin, and other tissue. Malonic acid is a strong irritant and is slightly toxic. When dissolved in water it is a strong acid and is corrosive to eyes, skin, and the respiratory tract. Potassium bromate is an oxidizer and presents a fire risk in contact with organic materials. It is a strong irritant and moderately toxic. This reaction generates a small amount of elemental bromine in aqueous solution—as noted by the appearance and disappearance of the orange color. Bromine water is toxic by inhalation and ingestion and is a skin irritant. Provide adequate ventilation. Wear chemical splash goggles, chemical-resistant gloves, and a chemical-resistant apron. Consult current Material Safety Data Sheets for additional safety, handling, and disposal information.

Preparation

To save time, measure or weigh out the required amounts of each reagent before the demonstration: 400 mL of 1.5 M sulfuric acid solution in a 500-mL graduated cylinder, and 4.5 g of malonic acid, 4.0 g of potassium bromate, and 1.3 g of manganous sulfate in separate, clearly labeled weighing dishes or small beakers.

Procedure

1. Measure 400 mL of 1.5 M sulfuric acid solution into a 1-L beaker containing a magnetic stirring bar.

2. Place the beaker on a magnetic stirrer and adjust the speed of the stirrer to ensure thorough and continuous mixing of the solution.

Demonstrations

3. Transfer 4.5 g of malonic acid to the beaker and stir until it is completely dissolved.

4. Add 4.0 g of potassium bromate to the resulting colorless solution and stir until dissolved.

5. Transfer 1.3 g of manganous sulfate to the beaker and continue stirring. Observe the solution as it turns amber-orange in color.

6. Within about 90 seconds the solution will revert to its original colorless state. This cycle will repeat itself approximately every 30 seconds for up to 30 minutes. The period between color oscillations will gradually increase as the reaction continues.

Disposal

Consult your *Flinn Scientific Catalog/Reference Manual* for general guidelines and specific procedures governing the disposal of laboratory wastes. The reaction mixture can be neutralized with sodium carbonate and flushed down the drain with excess water according to Flinn Suggested Disposal Method #24a.

Tips

- The equipment and glassware used in this demonstration should be clean and free of chloride ion contamination. Rinse all equipment and glassware in distilled water before use. Even small amounts of chloride ion will inhibit the oscillating chemical reaction.

- The use of a magnetic stirrer is highly recommended and will produce the best results.

- The color change becomes more distinct as the reaction proceeds; the colorless phase may initially have a bit of a rose tint.

- This is the simplest of the oscillating chemical reactions to perform, in terms of both the number of chemicals involved and the relative ease of interpreting the unusual reaction behavior. Use the demonstration as an alternative to the hypothetical textbook treatment of reaction pathways.

- A good way to initiate class discussion of this unusual chemical reaction is to emphasize the action of the Mn(II) reagent—when the the chemicals are mixed as described in steps 1–4, before Mn^{2+} ions are added, no reaction occurs. This is the action of a catalyst. How does a catalyst affect the rate and mechanism of a reaction?

Discussion

The unusual observations in this demonstration seem to contradict traditional chemical wisdom. Experience suggests that if a chemical system is left undisturbed, reactants should disappear and products should appear in a continuous manner until equilibrium is reached.

Oscillating chemical reactions appear to violate this wisdom, as an intermediate product (color) appears, disappears, and then reappears in a cycle that repeats itself many times. Several factors have been found to be important in setting up conditions for an oscillating chemical reaction to take place. First of all, the concentrations of reactants must be far away from their equilibrium values. Secondly, there must be two competing reaction pathways available for the overall reaction to take place. If one of the pathways produces an intermediate, while the other pathway consumes it, then the concentration of that intermediate will act like a trigger and the overall reaction will switch back and forth from one pathway to another.

Teacher Notes

Demonstrations

Teacher Notes

This oscillating chemical reaction demonstrates a modified Belousov-Zhabotinsky (BZ) reaction, manganese(II)-catalyzed oxidation of malonic acid by bromate ions. The overall redox reaction that takes place is shown in Equation 1. Oxidation of malonic acid to carbon dioxide, formic acid, and water is accompanied by reduction of bromate ion to bromide ion. The reaction requires the presence of Mn^{2+} catalyst.

$$CH_2(CO_2H)_2(aq) + BrO_3^-(aq) \xrightarrow{Mn^{2+}} HCO_2H(aq) + 2CO_2(g) + H_2O(l) + Br^-(aq)$$

The balanced equation does not provide any insight into how the reaction actually takes place. All of the reagents (both reactants and products) are colorless and cannot be responsible for the orange color observed. In order to understand the observations it is nesessary to recall the idea of a catalyst and to propose the existence of at least one intermediate.

Since Mn^{2+} ions are required for the reaction to start, but are neither a reactant nor a product, they must be acting as a catalyst. Assume that the catalyst allows the reaction to occur via a pathway that involves the formation and subsequent reaction of at least one reaction intermediate. This assumption explains two key observations: (1) the role of manganese ion in jump-starting the reaction, and (2) the appearance followed by disappearance of the orange color. The orange color is not associated with any of the reactants or products, and so it must arise from an intermediate. The characteristic orange color of bromine is well-known. Bromine may be the missing intermediate. But why does the orange color appear and disappear in a periodic fashion?

The key to understanding the repeated appearance and disappearance of bromine is a second assumption—there are actually two competing reaction pathways for the reduction of bromate ions. The observed oscillation reflects which pathway prevails under different conditions.

The two competing pathways illustrate two different ways in which electrons can be transferred in an oxidation–reduction process. In Process A, electrons are exchanged one electron at a time and Mn^{2+} ions are converted to Mn^{3+} ions as they reduce bromate to bromine (see Equation A1 on page 68). Bromine reacts with malonic acid via Equation B2, and bromide is formed as an overall product. In Process B, however, bromide ions appear as a reactant, transferring electrons two at a time as they reduce bromate ions to bromine (Equation B1). The limiting factor that determines whether Process A or Process B predominates is the bromide ion concentration. Bromide ions are not only a product of the overall reaction but also a reactant in Process B.

The demonstration begins with Process A, because bromide ions are not added to the reaction mixture. Process B takes over when the concentration of bromide ions, formed as a product in reaction A2, rises above a certain critical level. Oscillations are observed because as Process B continues, it depletes the bromide ion concentration—conditions that favor Process A once again. As the reaction switches between Process A and Process B, triggered by changes in the bromide ion concentration, the concentrations of other species in solution oscillate as well These concentration changes explain the oscillating color changes.

Demonstrations

Process A

$$2BrO_3^- + 12H^+ + 10Mn^{2+} \rightarrow Br_2 + 10Mn^{3+} + 6H_2O \quad \textit{Equation A1}$$

Bromate ions are reduced by manganese(II) ions to produce bromine through the redox reaction shown in Equation A1. Process A produces Mn(III) ions and Br_2. Both of these species react at least in part to oxidize the malonic acid (see Equation B2) and the bromomalonic acid (see Equation A2) to form bromide ions. As the concentration of bromide ions increases, the rate of Equation B1 increases until Process B begins to dominate.

$$BrCH(CO_2H)_2 + 4Mn^{3+} + 2H_2O \rightarrow Br^- + 4Mn^{2+} + HCO_2H + 2CO_2 + 5H^+$$

Equation A2

Process B

$$BrO_3^- + 5Br^- + 6H^+ \rightarrow 3Br_2 + 3H_2O \quad \textit{Equation B1}$$

Bromate ions are reduced by bromide ions through a series of oxygen transfers (two-electron reductions) summarized in Equation B1. The orange color which develops is caused by the production of elemental bromine. The color soon disappears as the bromine reacts with malonic acid as shown in Equation B2.

$$Br_2 + CH_2(CO_2H)_2 \rightarrow BrCH(CO_2H)_2 + Br^- + H^+ \quad \textit{Equation B2}$$

Process B results in an overall decline in the bromide ion concentration and, once the necessary intermediates are generated and most of the bromide ions are consumed, the rate becomes negligible and Process A again takes over.

Demonstrations

Teacher Notes

Sudsy Kinetics
An "Old Foamey" Demonstration

Introduction

Teach kinetics concepts in a new and sudsy way! This demonstration provides an interesting twist on the classic "Old Foamey" reaction. Not only will your students be amazed at the sudsy eruption—they will learn kinetics concepts along the way.

Concepts

- Kinetics
- Decomposition reaction
- Reaction intermediate
- Catalyst

Materials

Hydrogen peroxide, H_2O_2, 30%, 70 mL
Hydrogen peroxide, H_2O_2, 10%, 20 mL
Hydrogen peroxide, H_2O_2, 3%, 20 mL
Alconox® detergent, 20–25 g
Sodium iodide solution, NaI, 2 M, 25–30 mL
Wood splint
Graduated cylinders, 10-mL, 3
Graduated cylinders, 250-mL, 3
Graduated cylinders, 500-mL, 3
Large, plastic demonstration tray
Lighter or matches
Spoon or scoop

Safety Precautions

Hydrogen peroxide solution is a strong oxidizing agent; it is severely corrosive to the skin, eyes, and respiratory tract, and is a dangerous fire and explosion risk. Do not heat this substance. Sodium iodide solution is slightly toxic by ingestion. Do not stand over the reaction; steam and oxygen are produced quickly. Wear chemical splash goggles, chemical-resistant gloves, and a chemical-resistant apron. Please consult current Material Safety Data Sheets for additional safety, handling, and disposal information.

Procedure

Part A. Effect of Concentration on the Rate of a Reaction

1. Place three 250-mL graduated cylinders on a large, plastic demonstration tray.

2. Add 20 mL of 30% hydrogen peroxide to the first cylinder, 20 mL of 10% hydrogen peroxide to the second cylinder, and 20 mL of 3% hydrogen peroxide to the third cylinder.

3. Add 1 small scoop (3–4 g) of Alconox® to each cylinder and swirl to dissolve.

4. Measure 5 mL of 2 M sodium iodide solution into each of three 10-mL graduated cylinders. Ask students to predict how fast each of the hydrogen peroxide solutions will react.

5. Ask for three student volunteers—make sure they are wearing chemical splash goggles and warn them to step back as soon as they have poured the catalyst. Have the students simultaneously pour the sodium iodide solution into the three cylinders. *(White foam will cascade out of the cylinder containing 30% peroxide, will slowly rise from the 10% peroxide, and will be barely noticeable in the cylinder containing 3% peroxide.)*

Alconox detergent may be replaced with a liquid detergent (e.g., Ivory or Dove). The suds will not be as white or thick, but the demonstration still works well. Food coloring can also be added to the mixture (add with the soap) for special effects.

Demonstrations

Part B. Old Foamey—Observing a Reaction Intermediate and Products

1. Place a 500-mL graduated cylinder on a large, plastic demonstration tray.

2. Measure 20 mL of 30% hydrogen peroxide and add it to the cylinder.

3. Add 1 small scoop (3–4 g) of solid Alconox® detergent to the cylinder and swirl the mixture to dissolve the detergent.

4. Measure 5 mL of 2 M sodium iodide solution and, quickly but carefully, pour this into the cylinder. In a few seconds, copious amounts of white foam will be produced. Observe closely at the beginning of the reaction. *(A ring of brown foam is produced at first but then turns white before it erupts out of the cylinder. The brown color is due to iodine produced by the extreme oxidizing ability of the 30% hydrogen peroxide.)*

5. Observe the steam rising from the foam—the decomposition reaction is very exothermic.

6. Light a wood splint and blow out the flame. Insert the glowing wood splint into the foam. *(The wood splint will reignite in the foam. The gas that produces the foam is pure oxygen. Take the glowing splint out of the foam, reinsert it, and watch it reignite again. This can be repeated numerous times.)*

Part C. Compare the Rate of the Reaction versus its Stoichiometry

1. Place two 500-mL graduated cylinders in the center of a large, plastic demonstration tray.

2. Carefully measure 15 mL of 30% hydrogen peroxide and add it to the first cylinder.

3. Measure a second 15 mL portion of 30% hydrogen peroxide and add it to the second cylinder. To this second cylinder, also add 30 mL of tap water.

4. Add a scoop (3–4 g) of Alconox® detergent to each solution. Swirl to dissolve the detergent.

5. Ask for two student volunteers—make sure they are wearing goggles and warn them to step back as soon as they have poured the catalyst. Have them add 4 mL of 2 M sodium iodide solution to each of the two cylinders. *(White foam will immediately rise up out of each cylinder, with the more concentrated mixture reacting more rapidly. The diluted solution will eventually produce the same amount of foam as the first cylinder since an equal number of moles of hydrogen peroxide was used in both cases.)*

Disposal

Please consult your current *Flinn Scientific Catalog/Reference Manual* for general guidelines and specific procedures governing the disposal of laboratory wastes. The foam and any solutions left in the graduated cylinders or on the plastic tray may be rinsed down the drain with excess water according to Flinn Suggested Disposal Method #26b.

Tips

- Each part of the demonstration is designed to teach a different concept. Part A illustrates how the rate of the reaction depends on the concentration of hydrogen peroxide. Part B shows the reactions that are occurring, the brown iodine intermediate, the production of heat, and the formation of water and oxygen gas. Finally, Part C demonstrates the stoichiometry of the decomposition reaction.

Teacher Notes

If a large, plastic demonstration tray is not available, tape a plastic bag over the demonstration area or perform the demonstration in a laboratory sink.

Demonstrations

Teacher Notes

- The decomposition reaction of 30% hydrogen peroxide is highly exothermic and the graduated cylinder will become very hot. Allow it to cool before handling.

Discussion

Hydrogen peroxide decomposes to produce oxygen and water according to the following balanced chemical equation.

$$2H_2O_2(aq) \rightarrow 2H_2O(l) + O_2(g) + \text{heat}$$

The reaction is quite slow in the absence of a catalyst. Iodide ions, manganese metal, manganese dioxide, ferric ions, and many other substances such as yeast and blood will act as catalysts for the decomposition of hydrogen peroxide. A *catalyst* is a substance that, when added to a reaction mixture, participates in the reaction and speeds it up, but is not itself consumed in the reaction. Potassium iodide is used as a catalyst in this demonstration. The following reaction pathway has been proposed.

Step 1: Hydrogen peroxide and iodide spontaneously form a brown intermediate (I_3^-). Very little foam is formed in the initial stages of the reaction. Formation of I_3^- shows that the catalyst participates in the reaction pathway.

$$H_2O_2(aq) + 3I^-(aq) \rightarrow 2OH^-(aq) + I_3^-(aq)$$

Step 2: The intermediates react with additional hydrogen peroxide, resulting in the disappearance of the brown color and the production of copious amounts of foam containing oxygen gas. Iodide ions are regenerated—the catalyst is not consumed in the reaction.

$$H_2O_2(aq) + I_3^-(aq) + 2OH^-(aq) \rightarrow 2H_2O(l) + 3I^-(aq) + O_2(g)$$

Overall Reaction: Combining Steps 1 and 2 gives the overall reaction equation.

$$2H_2O_2(aq) \rightarrow 2H_2O(l) + O_2(g) + \text{heat}$$

The overall reaction is exothermic—the enthalpy or heat of reaction is negative. The free energy of the reaction, which takes into account not only the enthalpy but also the entropy of the reaction, is also negative, indicating that the reaction is spontaneous. If the reaction is spontaneous, then why is a catalyst needed? The catalyst causes the reaction to occur at a reasonable rate—without it, the reaction would occur, but so slowly that it would not be observable.

The fact that a spontaneous reaction nevertheless requires a catalyst points out a common misconception about the meaning of "spontaneous." In thermodynamics, a spontaneous reaction is one that will occur without outside intervention. It does not mean that the reaction is fast!

The following calculations show the maximum amount of oxygen that can be produced in Part C starting with 15 mL of 30% hydrogen peroxide.

Mass of 30% H_2O_2 (d = 1.11g/mL) = 15 mL × 1.11 g/mL = 16.7 g solution

$$\text{Moles of } H_2O_2 = 16.7 \text{ g} \times \frac{30 \text{ g } H_2O_2}{100 \text{ g solution}} \times \frac{1 \text{ mole}}{34.02 \text{ g}} = 0.15 \text{ moles } H_2O_2$$

$$\text{Moles of } O_2 = 0.15 \text{ moles } H_2O_2 \times \frac{1 \text{ mole } O_2}{2 \text{ moles } H_2O_2} = 0.075 \text{ moles } O_2$$

Volume of O_2 at 100 °C = (.075 moles)(.0821 L·atm/mole·K)(373 K)/1 atm = 2.3 L O_2

Sudsy Kinetics

Demonstrations

The Pink Catalyst

Teacher Notes

Introduction

Hydrogen peroxide is added to a hot solution of potassium sodium tartrate. Nothing much happens—the solution is colorless and a few bubbles are observed. Then add a special pink catalyst solution. Immediately the solution turns green and begins to bubble vigorously and froth over. When the bubbling subsides, the original pink color of the catalyst returns. What is a catalyst and how does it work?

Concepts

- Kinetics
- Catalyst

Materials

Cobalt chloride, $CoCl_2 \cdot 6H_2O$, 4–5 g

Hydrogen peroxide solution, H_2O_2, 6%, 40 mL

Potassium sodium tartrate, $KNaC_4H_4O_6 \cdot 4H_2O$, 0.2 M, 100 mL

Beaker, 50-mL

Beaker, 600-mL or 1-L

Distilled water

Graduated cylinder, 100-mL

Hot plate

Spatula or scoop

Stirring rod or magnetic stirrer

Thermometer

Safety Precautions

Cobalt chloride is moderately toxic by ingestion and causes blood damage. Hydrogen peroxide is a strong oxidizer and a skin and eye irritant. Avoid contact of all chemicals with eyes and skin. Wear chemical-resistant goggles, chemical-resistant gloves, and a chemical-resistant apron. Please review current Material Safety Data Sheets for additional safety, handling, and disposal information.

Preparation

Prepare 0.2 M potassium sodium tartrate solution by dissolving 6 g of $KNaC_4H_4O_6 \cdot 4H_2O$ in 100 mL of distilled water.

Procedure

1. Using a graduated cylinder, measure 100 mL of 0.2 M potassium sodium tartrate solution and pour it into a 600-mL or 1-L beaker.

2. Place the beaker on a hot plate at a medium setting and slowly warm the solution to 70 °C.

3. While waiting for the temperature of the solution to increase, dissolve one scoop (4–5 g) of cobalt chloride in a minimal amount (2–3 mL) of distilled water. Show this solution to the class so that the students can note the pink color of the catalyst.

4. When the temperature of the potassium sodium tartrate solution reaches 70 °C, carefully add 40 mL of 6% hydrogen peroxide solution. Is there any evidence of a chemical reaction? *(There may be a few bubbles here and there.)*

5. Add the pink catalyst solution to the reaction mixture and stir continuously.

Demonstrations

Teacher Notes

6. Observe the rate and the progress of the resulting chemical reaction. *(The solution immediately turns green and vigorous bubbling ensues. The mixture begins to froth and foam, then just as suddenly subsides. When the rate of bubbling diminishes, the green color disappears and the original pink color of the catalyst solution returns.)*

Disposal

Consult your current *Flinn Scientific Catalog/Reference Manual* for general guidelines and specific procedures governing the disposal of laboratory waste. The final cobalt-containing solution may be disposed of according to Flinn Suggested Disposal Method #27d.

Tips

- The demonstration may be extended to investigate the effect of temperature on the reaction rate. (Even catalyzed reactions are affected by the reaction temperature.) The reaction may be timed at various temperatures. Suggested temperatures and their corresponding reaction times are: 50 °C—200 seconds, 60 °C—90 seconds, and 70 °C—40 seconds.

- Begin timing the reaction upon addition of the pink catalyst solution. Complete the timing after the vigorous reaction subsides and the original pink color of the cobalt chloride solution has returned. Using a hot plate–magnetic stirrer combination is strongly recommended if you are going to measure reaction times. Otherwise, it will be necessary to continuously stir the reaction mixture.

Discussion

The reaction of tartrate ions with hydrogen peroxide is an example of an oxidation–reduction reaction. Hydrogen peroxide is a strong oxidizing agent, resulting in the complete oxidation of tartrate ions to give carbon dioxide and water (Equation 1). The extent of this oxidative decomposition reaction is evident by the production of carbon dioxide gas.

$$C_4H_4O_6^{2-} + 5H_2O_2 \rightarrow 4CO_2 + 6H_2O + 2OH^- \qquad \textit{Equation 1}$$

In the absence of a catalyst, the decomposition reaction, although thermodynamically favorable, is kinetically very slow. Thus, even at 75 °C, the reaction occurs at a barely noticeable rate.

To speed up the reaction, a catalyst must be used. Cobalt ions are known to catalyze the decomposition of hydrogen peroxide. The action of the cobalt catalyst can be followed by observing the color changes of the solution over the course of the reaction. The solution starts out pink, the color of the cobalt(II) aquo complex $[Co(H_2O)_6^{2+}]$. The mixture then quickly turns green, indicating the formation of a cobalt(III) complex. The rapid production of gas bubbles due to oxidation of the tartrate ions occurs almost immediately after the green color has been observed. As the tartrate ions are consumed in the reaction and the amount of gas production subsides, the color of the solution returns to the original pink color of the cobalt(II) catalyst. The following steps in the reaction pathway have been proposed based on the results of kinetic and mechanistic experiments.

The first step is the formation of a Co(II)-tartrate coordination compound. This is followed by oxidation of the pink Co(II) complex to a green Co(III) complex. The Co(III) complex is most likely a binuclear cobalt containing several tartrate ions. It is this unknown complex that is thought to be the actual catalyst in the oxidation reaction of tartrate ions by hydro-

The Pink Catalyst

Demonstrations

gen peroxide to give carbon dioxide and water. In the course of oxidation of the tartrate ions, the Co(III) complex ion is reduced back to Co(II). When all the tartrate has been consumed, the color of the solution reverts back to pink, indicating that only Co(II) ions are present in solution at the end of the reaction. Since the color of the solution is green throughout the reaction, most of the cobalt must be in the form of Co(III) ions during this time. This suggests that the first step, oxidation of Co(II) to Co(III) ions, is very fast compared to the second step, oxidation of the tartrate ions and reduction of the Co(III) complex.

Cobalt chloride is a catalyst because it is not consumed during the course of the reaction and it greatly speeds up the reaction. The cobalt chloride could also be isolated from the reaction mixture and reused in future reactions.

A Simple Catalyst Demo

Try the following activity to demonstrate the properties of a catalyst. All that is needed are a new box of chalk, a shoe box, and a heavy weight or rock.

Remove half the chalk pieces from the chalk box and place them in a shoe box. Each piece of chalk represents a long molecule, such as a high-molecular weight hydrocarbon in oil. The reaction we want to model involves breaking this long molecule into shorter molecules, similar to the reaction that occurs in petroleum refining to obtain gasoline from crude oil.

Place the chalk in a shoe box, close the shoe box, and shake it vigorously for 15 seconds. Open the box and remove the "products." Have the students examine the products and observe how well the reaction has occurred.

Take the rest of the new chalk pieces and place them in the shoe box, along with a special "catalyst," which may be a rock, weight, or any other unbreakable, heavy object. Close the box and shake it vigorously for 15 seconds. Open the box and remove the catalyst—show that the catalyst may be recovered unchanged from the reaction mixture. Then remove the products and compare the products from the first, uncatalyzed reaction with those from the second, catalyzed reaction. Which reaction occurred faster? What would happen if the first reaction were shaken for a longer period of time? Would the products of the two reactions look similar? Does the catalyst change the reaction products, or just speed the reaction up? Has the catalyst changed? Can the catalyst be used again and again?

Teacher Notes

Teacher Notes

The Floating Catalyst
An Enzyme Reaction Demonstration

Introduction

Almost all of the chemical reactions that take place in living organisms are catalyzed by enzymes—nature's catalysts. Enzymes are highly active catalysts. A typical enzyme, for example, may make a chemical reaction occur about one million times faster than it would in the absence of a catalyst. In this demonstration, we will look at the reaction of catalase, one of the most active known enzymes. Catalase catalyzes the decomposition of hydrogen peroxide in both plant and animal tissues.

Concepts

- Catalyst
- Enzyme
- Reaction rate
- Concentration

Materials

Catalase solution, 0.01%, 50 mL*
Hydrogen peroxide solution, H_2O_2, 3%, 1 L
Distilled or deionized water
Ice
Beef liver or potato (optional)
Beakers, 600-mL, 4
Evaporating dish or Petri dish
Filter paper, 5.5-cm diameter
Forceps or tongs
Graduated cylinders, 100- and 500-mL
Paper towels
Stirring rod
Stopwatch or timer

*Prepare the enzyme solution fresh before use and store over ice until needed. See the *Tips* section for instructions to prepare the enzyme solution and for alternative methods of extracting the enzyme from living tissue.

Safety Precautions

Hydrogen peroxide is a strong oxidizing agent and may be irritating to eyes and skin. Avoid contact of all chemicals with eyes and skin. Wear chemical splash goggles and chemical-resistant gloves and apron. Please review current Material Safety Data Sheets for additional safety, handling, and disposal information.

Procedure

1. Label four 600-mL beakers A–D.

2. Prepare a series of hydrogen peroxide solutions at different concentrations. Using Table 1 as a guide, measure and add the appropriate amounts of 3% hydrogen peroxide and distilled water into each beaker. Use a 100-mL or 500-mL graduated cylinder to measure the liquid volumes, as needed. There should be 500 mL of solution in each beaker.

The Floating Catalyst

Demonstrations

Teacher Notes

Table 1

Beaker	A	B	C	D
H$_2$O$_2$, mL	330 mL	170 mL	85 mL	40 mL
Distilled Water, mL	170 mL	330 mL	415 mL	460 mL
Concentration H$_2$O$_2$, %	2%	1%	0.5%	0.24%

3. Stir the solution in each beaker with a stirring rod to ensure thorough mixing. Rinse and pat dry the stirring rod between solutions.

4. Pour 20–30 mL of the catalase solution into a wide shallow dish such as a Petri dish or an evaporating dish.

5. Immerse four pieces of filter paper in the catalase solution and soak the filter paper for 2–3 minutes.

6. Using forceps or tongs, remove the filter paper circles from the catalase solution and blot them dry on a paper towel.

7. Using long forceps or crucible tongs, submerge one piece of filter paper in the bottom of the hydrogen peroxide solution in Beaker A. Release the filter paper and immediately start timing. *(The solution will start bubbling at the surface of the filter paper and the filter paper will gradually float to the surface of the hydrogen peroxide solution.)*

8. Measure and record the time in seconds when the center of the filter paper touches the top surface of the solution.

9. Repeat steps 7 and 8 three more times using Beakers B, C, D.

10. *(Optional)* Repeat the demonstration using fresh pieces of catalase-soaked filter paper and average the reaction times at each concentration. It is not necessary to prepare fresh hydrogen peroxide solutions—these solutions may be reused several times.

11. Compare the reaction times for the four different concentrations of hydrogen peroxide. How does the concentration of hydrogen peroxide affect the rate of the catalase reaction?

12. *(Optional)* Repeat the demonstration using a natural extract of the catalase enzyme from potato or beef liver. (See the *Tips* section.)

Tips

- The time required for the filter paper to rise to the surface depends on the activity of the catalase solution, which is related to both its concentration and the purity or activity of the enzyme itself. The shelf life of many commercially available enzymes is poor. Store enzymes in a refrigerator and use them within one year of purchase.

- To prepare a 0.01% catalase solution, first prepare 100 mL of a 0.1% solution by dissolving 0.10 g of catalase in 100 mL of distilled or deionized water. Dilute this solution tenfold by adding 90 mL of distilled water to 10 mL of the 0.1% solution to make 100 mL of 0.01% catalase. Test the activity of the enzyme solution in 2% hydrogen peroxide before performing the demonstration. Adjust the concentration as needed to obtain convenient "floating" times (neither too fast nor too slow).

Demonstrations

Teacher Notes

- The enzyme catalase may be extracted from living tissue. Cut small sections (about 1 cm^3) of potato or beef liver, mash or grind them, and soak the pulp in 50 mL of ice-cold distilled water for 10 minutes. Strain the extract through cheesecloth and test its activity in 2% hydrogen peroxide. Dilute the extract, if necessary, to obtain convenient reaction times.

Discussion

Decomposition of hydrogen peroxide to produce water and oxygen gas (Equation 1) is thermodynamically favorable but kinetically very slow in the absence of a catalyst.

$$2H_2O_2(aq) \rightarrow 2H_2O(l) + O_2(g) \qquad \textit{Equation 1}$$

In nature, this reaction is catalyzed by the enzyme catalase. This is an important reaction within cells. Hydrogen peroxide is generated as a byproduct of metabolic processes and the enzyme catalase prevents the accumulation of dangerous levels of this toxic chemical.

The rate of the catalase reaction can be determined by measuring the time required for the enzyme-soaked filter paper disk to rise to the surface in a solution of hydrogen peroxide. Oxygen bubbles form on the filter paper and cause it to float. The rate of the reaction is inversely related to the reaction time. Table 2 gives some typical results for this demonstration.

Table 2

Beaker	A	B	C	D
Concentration of H_2O_2	2%	1%	0.5%	0.24%
Average Reaction Time	8 sec	11 sec	18 sec	31 sec
Average Rate	0.13 sec^{-1}	0.09 sec^{-1}	0.06 sec^{-1}	0.03 sec^{-1}

At low concentrations of hydrogen peroxide, the rate of the reaction increases almost linearly as the concentration increases. At higher concentrations of hydrogen peroxide, the enzyme-catalyzed reaction behaves differently than a typical chemical reaction—the rate increase becomes more gradual. Eventually, the rate of the enzyme-catalyzed reaction would be expected to "level off" or reach a maximum value. A plot of the rate of the catalase reaction versus hydrogen peroxide concentration is shown below. The shape of the curve is characteristic of an enzyme-catalyzed reaction.

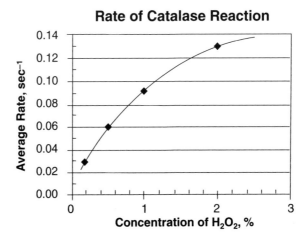

The Floating Catalyst

Safety and Disposal Guidelines

Safety Guidelines

Teachers owe their students a duty of care to protect them from harm and to take reasonable precautions to prevent accidents from occurring. A teacher's duty of care includes the following:

- Supervising students in the classroom.
- Providing adequate instructions for students to perform the tasks required of them.
- Warning students of the possible dangers involved in performing the activity.
- Providing safe facilities and equipment for the performance of the activity.
- Maintaining laboratory equipment in proper working order.

Safety Contract

The first step in creating a safe laboratory environment is to develop a safety contract that describes the rules of the laboratory for your students. Before a student ever sets foot in a laboratory, the safety contract should be reviewed and then signed by the student and a parent or guardian. Please contact Flinn Scientific at 800-452-1261 or visit the Flinn Website at www.flinnsci.com to request a free copy of the Flinn Scientific Safety Contract.

To fulfill your duty of care, observe the following guidelines:

1. **Be prepared.** Practice all experiments and demonstrations beforehand. Never perform a lab activity if you have not tested it, if you do not understand it, or if you do not have the resources to perform it safely.

2. **Set a good example.** The teacher is the most visible and important role model. Wear your safety goggles whenever you are working in the lab, even (or especially) when class is not in session. Students learn from your good example—whether you are preparing reagents, testing a procedure, or performing a demonstration.

3. **Maintain a safe lab environment.** Provide high-quality goggles that offer adequate protection and are comfortable to wear. Make sure there is proper safety equipment in the laboratory and that it is maintained in good working order. Inspect all safety equipment on a regular basis to ensure its readiness.

4. **Start with safety.** Incorporate safety into each laboratory exercise. Begin each lab period with a discussion of the properties of the chemicals or procedures used in the experiment and any special precautions—including goggle use—that must be observed. Pre-lab assignments are an ideal mechanism to ensure that students are prepared for lab and understand the safety precautions. Record all safety instruction in your lesson plan.

5. **Proper instruction.** Demonstrate new or unusual laboratory procedures before every activity. Instruct students on the safe way to handle chemicals, glassware, and equipment.

6. **Supervision.** Never leave students unattended—always provide adequate supervision. Work with school administrators to make sure that class size does not exceed the capacity of the room or your ability to maintain a safe lab environment. Be prepared and alert to what students are doing so that you can prevent accidents before they happen.

7. **Understand your resources.** Know yourself, your students, and your resources. Use discretion in choosing experiments and demonstrations that match your background and fit within the knowledge and skill level of your students and the resources of your classroom. You are the best judge of what will work or not. Do not perform any activities that you feel are unsafe, that you are uncomfortable performing, or that you do not have the proper equipment for.

Safety Precautions

Specific safety precautions have been written for every experiment and demonstration in this book. The safety information describes the hazardous nature of each chemical and the specific precautions that must be followed to avoid exposure or accidents. The safety section also alerts you to potential dangers in the procedure or techniques. Regardless of what lab program you use, it is important to maintain a library of current Material Safety Data Sheets for all chemicals in your inventory. Please consult current MSDS for additional safety, handling, and disposal information.

Disposal Procedures

The disposal procedures included in this book are based on the Suggested Laboratory Chemical Disposal Procedures found in the *Flinn Scientific Catalog/Reference Manual*. The disposal procedures are only suggestions—do not use these procedures without first consulting with your local government regulatory officials.

Many of the experiments and demonstrations produce small volumes of aqueous solutions that can be flushed down the drain with excess water. Do not use this procedure if your drains empty into groundwater through a septic system or into a storm sewer. Local regulations may be more strict on drain disposal than the practices suggested in this book and in the *Flinn Scientific Catalog/Reference Manual*. You must determine what types of disposal procedures are permitted in your area—contact your local authorities.

Any suggested disposal method that includes "discard in the trash" requires your active attention and involvement. Make sure that the material is no longer reactive, is placed in a suitable container (plastic bag or bottle), and is in accordance with local landfill regulations. Please do not inadvertently perform any extra "demonstrations" due to unpredictable chemical reactions occurring in your trash can. Think before you throw!

Finally, please read all the narratives before you attempt any Suggested Laboratory Chemical Disposal Procedure found in your current *Flinn Scientific Catalog/Reference Manual*.

Flinn Scientific is your most trusted and reliable source of reference, safety, and disposal information for all chemicals used in the high school science lab. To request a complimentary copy of the most recent *Flinn Scientific Catalog/Reference Manual,* call us at 800-452-1261 or visit our Web site at www.flinnsci.com.

National Science Education Standards

Experiments and Demonstrations

Content Standards	Introduction to Reaction Rates	Temperature and Reaction Rates	The Order of Reaction	Kinetics of Dye Fading	Determining a Rate Law	Iodine Clock Reaction	Now You See It— Now You Don't	Sudsy Kinetics	The Pink Catalyst	The Floating Catalyst
Unifying Concepts and Processes										
Systems, order, and organization	✓	✓	✓	✓	✓	✓	✓	✓	✓	✓
Evidence, models, and explanation	✓	✓	✓	✓	✓	✓	✓	✓	✓	✓
Constancy, change, and measurement	✓	✓	✓	✓	✓	✓	✓	✓	✓	✓
Evolution and equilibrium										
Form and function										
Science as Inquiry										
Identify questions and concepts that guide scientific investigation	✓	✓	✓	✓	✓	✓		✓		✓
Design and conduct scientific investigations	✓	✓	✓	✓	✓	✓	✓	✓	✓	✓
Use technology and mathematics to improve scientific investigations	✓	✓	✓	✓	✓	✓		✓		✓
Formulate and revise scientific explanations and models using logic and evidence	✓	✓	✓	✓	✓	✓	✓	✓	✓	✓
Recognize and analyze alternative explanations and models							✓		✓	
Communicate and defend a scientific argument										
Understanding scientific inquiry	✓	✓	✓	✓	✓	✓		✓		✓
Physical Science										
Structure of atoms										
Structure and properties of matter	✓	✓				✓				
Chemical reactions	✓	✓	✓	✓	✓	✓	✓	✓	✓	✓
Motions and forces										
Conservation of energy and the increase in disorder										
Interactions of energy and matter	✓	✓				✓		✓	✓	

National Science Education Standards

Content Standards (continued)

Experiments and Demonstrations

	Introduction to Reaction Rates	Temperature and Reaction Rates	The Order of Reaction	Kinetics of Dye Fading	Determining a Rate Law	Iodine Clock Reaction	Now You See It— Now You Don't	Sudsy Kinetics	The Pink Catalyst	The Floating Catalyst
Science and Technology										
Identify a problem or design an opportunity		✓								
Propose designs and choose between alternative solutions		✓								
Implement a proposed solution		✓								
Evaluate the solution and its consequences		✓								
Communicate the problem, process, and solution		✓								
Understand science and technology		✓						✓		
Science in Personal and Social Perspectives										
Personal and community health										
Population growth										
Natural resources										
Environmental quality										
Natural and human-induced hazards										
Science and technology in local, national, and global challenges										
History and Nature of Science										
Science as a human endeavor										
Nature of scientific knowledge	✓	✓	✓	✓	✓	✓	✓	✓	✓	✓
Historical perspectives										

Master Materials Guide

(for a class of 30 students working in pairs)

Experiments and Demonstrations

Chemicals	Flinn Scientific Catalog No.	Introduction to Reaction Rates	Temperature and Reaction Rates	The Order of Reaction	Kinetics of Dye Fading	Determining a Rate Law	Iodine Clock Reaction	Now You See It—Now You Don't	Sudsy Kinetics	The Pink Catalyst	The Floating Catalyst
Acetic acid, 6 M	A0186			3 mL							
Alconox® detergent	A0126								20–25 g		
Catalase	C0359										0.1 g
Cobalt chloride	C0225									4–5 g	
Copper wire, 18 gauge	C0148		6 m								
Dextrose, anhydrous	D0002	18									
Ethyl alcohol, anhydrous	E0007				20 mL						
Hydrochloric acid, 1 M	H0057		2 L			500 mL					
Hydrogen peroxide, 30%	H0037								77 mL		
Hydrogen peroxide solution, 6%	H0028								40 mL		
Hydrogen peroxide solution, 3%	H0009			23 mL					20 mL		1 L
Magnesium ribbon	M0139		4 m								
Malonic acid	M0091								5 g		
Manganous sulfate	M0030								2 g		
Methylene blue	M0072	1 g									
—or— Methylene blue solution, 1%	M0074	10 mL*									
Phenolphthalein solution, 1%	P0019				5 mL						
Potassium bromate	P0205								4 g		
Potassium hydroxide	P0058	59 g									
Potassium iodide	P0066			2 g							
Potassium iodide solution, 0.2 M	P0168					325 mL					
Potassium sodium tartrate	P0084									6 g	
Sodium acetate	S0344			2 g							
Sodium hydroxide, pellets	S0074				1 g						
Sodium iodide	S0083								15 g		
Sodium metabisulfite	S0317								4 g		
Sodium thiosulfate pentahydrate	S0114			1 g		75 g					
Spray starch	S0302			1							
Starch, soluble	S0122								4 g		
Sulfuric acid, conc., 12.5 M	S0380								48 mL		
Sulfuric acid solution, 0.1 M	S0419						10 mL				

* Dilute 1/10

Master Materials Guide

(for a class of 30 students working in pairs) — **Experiments and Demonstrations**

Item	Flinn Scientific Catalog No.	Introduction to Reaction Rates	Temperature and Reaction Rates	The Order of Reaction	Kinetics of Dye Fading	Determining a Rate Law	Iodine Clock Reaction	Now You See It—Now You Don't	Sudsy Kinetics	The Pink Catalyst	The Floating Catalyst
Glassware											
Beakers											
50-mL	GP1005		15		45					1	
150-mL	GP1015	20*									
250-mL	GP1020					6					
400-mL	GP1025		15–20*			6					
600-mL	GP1030									1	4
1-L	GP1040						1				
Graduated cylinders											
10-mL	GP2005	15					1		3		
50-mL	GP2015		15				1				
100-mL	GP2020						1			1	1
250-mL	GP2030								3		
500-mL	GP2030								3		1
Stirring rod	GP5075						1			1	1
Test tubes											
15 × 125 mm	GP6015	45									
18 × 150 mm	GP6067		90								
General Equipment and Miscellaneous											
Balance, centigram	OB2059						1	1			
Cassette box	AP1519		15								
Colorimeter sensor	TC1504			15							
Cotton swabs	AP1737			30	60						
Cuvettes with lids	AP9149			15							
Demo tray, large, plastic	AP5429								1		
Evaporating dish	AP1272										1
Filter paper, 5.5 cm diam.	AP8994										4
Forceps	AP8328										
—or—											
Tongs	AP1359										1
Hot plate	AP4674	3–5 *	5–6*				1			1	
Labeling pens	AP1297	15				15					

*Several groups may share beakers to make water baths at different temperatures.

Continued on next page

Master Materials Guide

(for a class of 30 students working in pairs)

Experiments and Demonstrations

General Equipment and Miscellaneous, cont'd.	Flinn Scientific Catalog No.	Introduction to Reaction Rates	Temperature and Reaction Rates	The Order of Reaction	Kinetics of Dye Fading	Determining a Rate Law	Iodine Clock Reaction	Now You See It— Now You Don't	Sudsy Kinetics	The Pink Catalyst	The Floating Catalyst
Labels, adhesive	AP5368			60							
Lab Pro™ Interface System	TC1500				15						
Lighter, butane	AP8960								1		
Logger Pro™ software	TC1421				1						
Magnetic stirrer	AP6067						1			1	
Magnetic stir bar	AP1090							1		1	
Pipets											
Beral-type, thin stem	AP1444	60									
Beral-type, microtip	AP1517		60								
Reaction plates, 6-well	AP1725				15						
Reaction strips, 12-well	AP1446			30							
Ruler, metric	AP6324	15	15								
Scoop	AP8338								1	1	
Stoppers											
Size 0	AP2222	45									
Size 3, one-hole	AP2303		30								
Stopwatch —or— Timer	P1572 AP8874	15	15	15		15	1				1
Syringe, 3-mL	AP1728				15						
Test tube rack	AP1319	15	15								
Thermometer, digital	AP6049	15	15		15		1			1	
Tissues or lens paper, lint-free	AP1141				15						
Toothpicks, plastic	AP1810			60							
Wash bottle	AP1668	15			15	15					
Water, distilled or deionized	W0007 W0001	✓		✓	✓	✓	✓			✓	✓
Weighing dishes, plastic	AP1278						3				
Wooden splints	AP4444								1		